Recent Trends in AI and Data Science

Editors

S. Vijayalakshmi
Department of Computer Science
School of Sciences
CHRIST (Deemed to be University)
Pune, Lavasa, India

Savithri M.
Department of Data Science
CHRIST University
Bengaluru, India

Lija Jacob
School of Sciences
CHRIST (Deemed to be University)
Pune, Lavasa, India

CRC Press
Taylor & Francis Group
Boca Raton London New York

CRC Press is an imprint of the
Taylor & Francis Group, an **informa** business
A SCIENCE PUBLISHERS BOOK

Cover credit: AI generated image provided by the second editor, Savithri M.

First edition published 2027
by CRC Press
2385 NW Executive Center Drive, Suite 320, Boca Raton FL 33431

and by CRC Press
4 Park Square, Milton Park, Abingdon, Oxon, OX14 4RN

CRC Press is an imprint of Taylor & Francis Group, LLC

Library of Congress Cataloging-in-Publication Data (applied for)

ISBN: 978-1-041-05293-7 (hbk)
ISBN: 978-1-041-06386-5 (pbk)
ISBN: 978-1-003-63517-8 (ebk)

DOI: 10.1201/9781003635178

Typeset in Times New Roman
by Prime Publishing Services

Preface

The rapid evolution of Artificial Intelligence (AI) and Data Science has redefined the technological landscape across industries. With their powerful ability to process massive datasets, recognize complex patterns, and derive actionable insights, these fields have become the cornerstone of modern innovation. Recent Trends in AI and Data Science provides an in-depth exploration of the latest developments, methodologies, and real-world applications in these dynamic domains.

This edited volume is an extended version of a few research articles from the "International Conference on Artificial Intelligence, Communication, IoT, Data Engineering and Security," which brought together academicians, researchers, and industry experts to share their innovative ideas and emerging research in these fields. The chapters included in this book were carefully selected and reviewed from the papers presented at the conference, representing the best contributions that highlight current advancements and practical implementations in AI and Data Science.

The book brings together diverse research contributions that demonstrate how AI-driven systems and data analytics are revolutionizing domains such as healthcare, education, agriculture, cybersecurity, finance, and blockchain technology. The chapters encompass a wide spectrum of studies—from predictive modeling and neural network architectures to blockchain frameworks and ethical considerations in intelligent systems. Each contribution reflects the growing synergy between computational intelligence and domain-specific problem-solving.

The book begins with analytical and predictive frameworks, such as SARIMAX-based IT Index Modeling, that exemplify how time-series forecasting and market indicators can guide strategic decisions. It then delves into AI-driven educational analysis using Existential Intelligence, followed by healthcare innovations including NeuroNET and CNN-based models for brain tumor and lung cancer detection, which highlight the precision and reliability of deep learning in medical imaging.

Subsequent chapters explore the social and ethical dimensions of AI, including sentiment analysis on social media, algorithmic strategies for copyright protection, and AI-powered FinTech adoption studies using structural modeling approaches. These works illustrate the transformative power of AI in human-centered applications and governance systems. Other contributions, such as robo-advisor ethics and compliance models, emphasize integrating responsible AI frameworks into financial technologies.

A notable section of this book focuses on blockchain technologies—ranging from employee lifecycle management to topology-aware Raft consensus mechanisms—showcasing how decentralized architectures enhance transparency, trust, and efficiency. The book also features innovations in data security, including multi-field encryption and AI-based cyberattack detection, underscoring the need for secure digital infrastructures in the era of intelligent automation.

The final part of the book addresses AI applications in agriculture and environmental modeling, such as paddy and plant disease detection using ResNet and GLCM, and 3D LiDAR data compression using graph partitioning. These chapters highlight how deep learning and data science are being applied to solve real-world problems beyond the digital sphere, fostering sustainability and precision agriculture.

Through these multidisciplinary perspectives, "Recent Trends in AI and Data Science" aims to serve as a valuable resource for researchers, academicians, and practitioners. The book emphasizes not only the technological depth of AI and data analytics but also their ethical, social, and environmental implications. As a collective outcome of the conference, it reflects a shared vision of advancing knowledge and promoting collaborative research in emerging areas of intelligent systems and data-driven innovation. By combining theoretical insights with practical frameworks, this volume provides readers with a holistic understanding of how intelligent systems continue to shape the future of computation, communication, and decision-making.

Contents

AI in Finance and Compliance

Security and Data Protection

AI in Agriculture and Environment

Blockchain and System Performance

Foundational AI and Machine Learning Models

Chapter 1

IT Index Predictive Modelling

A Deep Study With SARIMAX and Market Indicators

Abhijeet Birari,[1,*] *Harshal Salunkhe,*[2] *Prajakta Yawalkar*[3] and *Jitendrasinh Jamadar*[4]

1. Introduction

The economy's current state has been reflected in the stock market. Various factors will influence market fluctuations, including firm-specific data, industry dynamics, or economic policy. It is also heavily influenced by market rumours and investor psychology. It is hard to forecast the stock value due to the stock market's complexity, ambiguity, volatility, and uncertainty. Various tools have been developed using statistical techniques that help to anticipate prices with an acceptable level of accuracy, notwithstanding their difficulty.

Researchers in the field of stock market analysis primarily use the ARIMA model for analysis. This model is mainly used to predict future market prices by analysing past price behaviour and the price pattern in the past. This model is proven to be a more effective and reliable model in forecasting the stock market price in various investigations [1, 2].

One of the most complex and volatile system that affect a firm's broad and general indices of macroeconomy is the stock market of the firm. The stock market has a significant impact based on the various economic polices like changes in

[1] Christ (Deemed to be University), Pune, Lavasa Campus.
[2] Associate Professor, MIT WPU, Pune.
[3] Professor, D Y Patil B School, Pune.
[4] Associate Professor, Nath School of Business and Technology, Maharashtra.
Emails: harshal.salunkhe@mitwpu.ac.in, prajakta.yawalkar@christuniversity.in, jitendrajamadar@gmail.com
* Corresponding author: abhijeet.birari@christuniversity.in

DOI: 10.1201/9781003635178-2

interest rates, amount spent by the government, etc. Industry dynamics also play a major role in the stock market based on the demand for their product in the market, technical breakthroughs, and the competitors in the market. The other information that affects the stock market price is based on the unique features of the firm, such as a change in management, a new product launch, and reports on earnings.

Market volatility might be caused by the psychology of the investor, and the market complexity will increase based on the feelings of the investor, such as greed and fear, which have a major influence on trade decisions. The prediction of the stock market is mostly unpredictable because of rumours and hypothetical activity due to a confused environment. A well-known model that is considered the backbone of analysing time series data, specifically in the financial market, is the ARIMA model. The temporal dependencies and the seasonality of the stock price can be identified using this model. Autoregressive (AR) and the moving average (MA) are integrated with this model to obtain stationarity, such that it will pick the short-term pattern and the long-term patterns in the given data. Various studies on the prediction of the stock market using the ARIMA model by various researchers have proved the effectiveness and robustness of the model. The accuracy of the model in forecasting based on certain conditions shows that the ARIMA model is a valuable tool for trade analysis. The information technology (IT) industry is playing a vital role in the stock market globally.

Over the past few decades, the IT industry has improved its innovative applications and experienced tremendous growth in the advancement of technology. The development of digital technology, the growth of the Internet, and the development of ingenuity in software and hardware solutions have made this an integral part of the world economy. This impact on the stock market is substantial, given that IT companies are generally the drivers of market trends, have substantial capitalization, and influence nearly all other sectors. Given its importance, predictive analysis has focused on it. Researchers and analysts have developed dozens of predictive models to forecast the performance of the IT indices, which are indicative of the collective performance of the IT stocks. These are essential indices that are read to inform policymaking for investors and portfolio managers at one level, and for policymakers in the finance ministry and the State Bank in another. IT indices are essential in investment decisions, portfolio management, and policy formulation [3].

The current paper focuses on building predictive models using established techniques, such as ARIMA and its relative extensions, targeting an IT index for India. An extension of the traditional ARIMA model is SARIMAX, which includes exogenous variables, allowing for more comprehensive modelling of time series data. The influences on stock prices deriving from outside, such as economic indicators, industry-specific developments, market sentiment, etc., are captured. To make the model more reliable and robust, these variables are been incorporated in the model.

IT index is considered as one of the most important parts in the Indian market, it also includes the details of all the major IT company that are controlling the particular sector. These companies are mainly involved in the production of hardware, IT services, and application development. The impact of the stock market index mainly depends on the technological improvements, changes in the regulatory environment, economic policy, world economy, and also events based on firm-specific factors. To provide a valuable insight to the stakeholders, it is necessary to understand the multiple factors that influence the correct prediction. In this chapter, the ARIMA and SARIMAX models are discussed based on the prediction and forecasting of the IT index in India. In the proposed system, new variables are added, like moving average, daily price disparity, Average True Range, and the Relative Strength Index, so that it will help in improving the accuracy of the prediction. This model will be more useful for stock market price prediction, along with the practical implications for stakeholders. This chapter will show how ARIMA and SARIMAX can be efficiently used in capturing the IT index dynamically. The model will become a strong tool for financial analysis based on the results analysis and its impact.

The following equation represents the ARIMA model:

$$\left(1-\sum_{i=1}^{p}\varnothing_i L^i\right)(1-L)^d X_t = \mu + \left(1-\sum_{i=1}^{q}\varnothing_i L^i\right)\varepsilon_t$$

Where:

- p denotes the autoregressive term's order.
- d represents the differencing value used to fix the time series.
- q denotes the order of the moving average
- L represents the lag operator.
- Φ_i represents the parameter of AR and θi represents the parameters of MA [4].

The results of the trials and their analysis have unlocked the doors to research and experiments that benefit traders, investors, and policymakers [5].

2. Related Works

In this literature review, the various predictive models that are used in stock market prediction applications are highlighted. The various significant researches that are been done based on the ARIMA and SARIMAX model and its effectiveness are identified in the survey. In this, the authors [6] studied the S&P 500 by applying ARIMA and SARIMAX and concluded that the SARIMAX exceeded the ARIMA model when external factors predict stock prices. The authors [7] also worked on the Indian Stock Market and found similar observations for ARIMA and SARIMAX, with SARIMAX performing better in a few scenarios.

The authors [8] specifically worked on the NASDAQ, which brought the interesting result that the basic structure of SARIMAX would provide robust predictions, while the Long Short-Term Memory (LSTM) models captured nonlinear patterns in better performance terms. For the Dow Jones Industrial Average, the authors [9] showed that short-term forecasting could be done effectively with ARIMA, but it was not effective for long-term forecasting. The author [10] analysed the Tokyo Stock Exchange. He proved that though SARIMAX can capture seasonality effects in exchanges, neural networks will have the upper hand in forecasting the stock markets due to their accuracy. In the paper [11], which points out that ARIMA could effectively do its forecasting accurately for linear trends in the Shanghai Stock Exchange, but it failed when the market faced sudden shocks.

The authors [12] conducted the same study on KOSPI, establishing that SARIMAX did capture seasonal patterns, but GARCH was better as a volatility model. The authors [13] surveyed the Mexican stock market, using ARIMA as a benchmark. Indeed, he concluded that machine learning techniques can perform better in predicting stock. While the working of authors [14] on the FTSE 100, reported that SARIMAX captured seasonal patterns and the effect of exogenous factors well. The paper [15] focused on the Australian Stock Exchange, finding that it was established that the combination of ARIMA with time series decomposition techniques had fairly descriptive predictions. The authors [16] worked on the New York Stock Exchange and observed that although deep learning models yielded better results in comparison to the SARIMAX model, the SARIMAX model provided interpretability for the model. The authors [17] also added that ARIMA was a valuable tool for forecasts in the Brazilian Stock Market, especially for making short-term predictions in an emerging market.

In their paper, the authors [18] even found that it gives better predictions for the Canadian Stock Market by including external regressors in their model. In their paper [19], the authors analysed the European Stock Markets to conclude that, although the ARIMA model can capture any linear trends, this is not enough to investigate such markets without the corresponding volatility models thoroughly. Further in the paper [20], it is demonstrated that the seasonally augmented model outperformed the simple ARIMA in the South African Stock Exchange, which operates on seasonal patterns and external factors.

The studies described above collectively highlight the strengths and weaknesses of these models on different stock markets and, most importantly, the need to consider exogenous variables and exercise model selection guided by specific market conditions.

3. Methodology

This research aims to predict the shift of the IT Index with the ARIMA model. The following phases provide the research methodologies to attain this goal (Figure 1.1).

- For this study, the dataset has been carefully sourced from Yahoo Finance. It captured crucial variables, including date, open, high, low, close, and volume of the Indian IT index between January 1, 2017, and September 21, 2023. The application of a large dataset forms a strong basis for analysis and modelling and impacts the performance of the IT Index. IT Index's daily closing price is considered as critical variable which is ideal for market trend indicator and movement.
- The feature engineering method might extract seven new features from the raw data, which comprise:
  - high-low
  - open-close

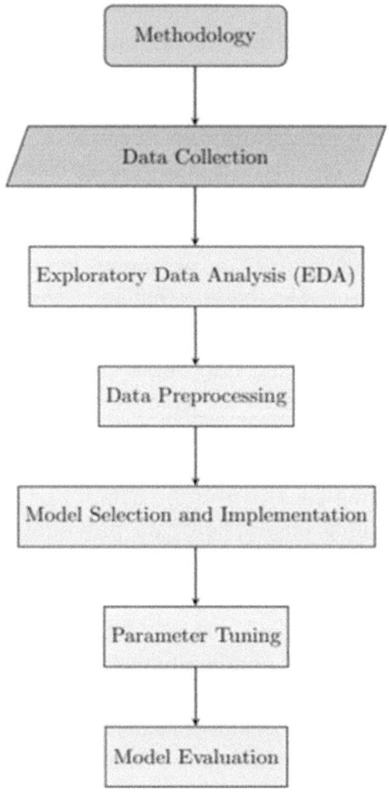

Figure 1.1 Methodology.

- 7-Day Moving Average (DMA)
- 14 DMA
- 21 DMA
- Relative Strength Index (RSI)
- Average True Range (ATR).

The above variables are recognised and selected founded on their significance and the likely impact on the closing price of the IT Index. The predictive capacity of the model is increased by feature engineering, which is considered one of the most important parts of data pre-processing.

To uncover the underlying trend in the market fluctuation, the difference between the high and the low value in a day, along with the opening index and closing index, is taken to identify the volatility in the daily market price interval, and the moving average is used to smooth the short-term fluctuations. The RSI and the ATR are other widely used technical indicators applied for measuring momentum and volatility, respectively, in a trading setting.

- Here, an initial exploratory data analysis has been conducted to detect any special structures or trends within the dataset. EDA is the process of analysing data sets to summarise their main characteristics using visual methods. Plots such as Line plots, histograms, and correlation matrices have been used to deduce preliminary information regarding the distribution, central tendency, and relationships among the variables. EDA plays a vital role in identifying and clarifying the deeper structure that helps in the selection of an appropriate technique used in modelling. Existence of seasonal patterns or trends in the closing price is determined using the line plots of closing price, and a histogram is used to show the distribution of returns. The features that affect the performance of the model are also examined with the correlation matrices.
- The given data is been divided into 80/20 ratio for training and testing. The training of the model is done on the training set and the testing is done on the other data set to check the performance of the model. The test reveals that the model is not overfitted with the training data but it provides a generalized with unseen data. By holding back part of the data for testing, a study can estimate how well a model predicts unseen new data, providing a more realistic appraisal of its performance.
- The reason the ARIMA model was used is due to the capability of the ARIMA model in modelling time-series data. ARIMA stands for autoregressive integrated moving-average, a commonly used statistical method for modelling and forecasting time-series data. The ensuing model analyses past price behaviour, diagnoses patterns, and forecasts future movements. Yet, the ARIMA does not allow external predictors to be accommodated; thus, the SARIMAX model has been applied in this study. In extending ARIMA to SARIMAX, the addition of exogenous variables includes external predictors alongside the ARIMA components to capture the further influence they might have on the target variable. The existence of exogenous variables in the model enables it to consider additional factors that would affect the closing prices of the IT Index for broader analysis. For example, some of these economic indicators include interest rates, industry-specific variables like technological advancements, etc., which may also be considered external predictors for an SARIMAX model to improve the forecasting power.
- This study identified, by the grid search technique, the best combination of parameters regarding ARIMA among several combinations of AR, I, and MA terms. Grid search is basically a very ordered way of maximizing hyperparameters. It checks on different parameter values to get the best combination possible to reduce errors on the validation set. This guarantees that a model is being developed in the finest possible manner so that it is trained with better accuracy and exactness. Practical model fine-tuning through hyperparameters improves performance since better parameter choice may lead to better quality predictions.
- The effectiveness of the model is assessed by comparing the predicted value from the test data set and the actual value that is achieved. Various key evaluation metrics like BIC, adjusted R2, SER, etc. are used in different combinations with

ARIMA model to identify the best-fitting of the model. BIC is used to identify the finite set from the various models. It is used to identify the best-fitting model by balancing the complexity and the goodness of the model. The model with lower BIC is preferable. Adjusted R2 specifies how well a model describes the dissimilarity of the target variable in relation to the number of predictors. The average distance of the observed values from the regression line is depicted by the SER, which in turn tells how well the model fits the data.

- Merging ARIMA and SARIMAX with strong feature engineering and optimization of hyperparameters, derive strong forecasting models for the IT Index. The study aimed to provide valuable insights into the performance of the IT Index by using advanced techniques and adequate data analysis. Implications drawn for investors, portfolio managers, and policymakers indicate that financial analysis and decision-making must incorporate external factors using advanced modelling techniques. It contributes to the extant body of knowledge regarding stock price prediction.

4. Result Analysis

The SARIMAX model has outperformed the classical ARIMA model due to the inclusion and selection of exogenous variables. The presence and role of the external predictor variables help discriminate between the models and analyse the significance of the variables in the SARIMAX model against the ARIMA model.

4.1 Model Significance

The SARIMAX model has been developed using the IT Index closing price as the dependent variable. The closing price captures much of the performance and trends in the market. The choice of the dependent variable is very critical. The closing price is considered the most crucial price during the trading day since it encompasses the net effect of all activities and sentiments of the market. The study also included several potential predictor variables to enhance the model's predictability. Each variable was based on its inherent tendency and capacity to influence the closing price of the IT Index to analyse and present as many aspects of market dynamics as possible. The following were some of the tested predictor variables:

Difference With the Maximum and Minimum Prices for the Day (H-L)

It reflects the daily range of prices and conveys the level of market volatility and the extent of intra-day price variations. A large difference between the high and low prices would signal high volatility, suggest market uncertainty, or indicate higher activity. In our study, the model tries to capture those daily market swings around the close price.

Open-Close (O-C): The Variance Between the Open and Close of a Day

This variable shows what happened from the opening until the session closes that day and explains the underlying sentiment in the market about that day. If the difference is positive, it means the market gained strength; if the difference is negative, it means a decline. The O-C measure captures the intraday momentum and market direction, providing insight into how the day's trading dynamics will pan out.

7-Day Moving Average (7DMA)

The 7DMA smoothens short-run fluctuations while highlighting the underlying trend over a week. It is a standard, extensively used technical indicator that helps identify the market's short-term direction. By taking an average of closing prices over seven days, the variable smooths the series and decreases the noise from daily price variations, giving a good view of the market's short-term trend.

14-Day Moving Average (14DMA)

It encompasses a two-week smoothing period, giving a medium-term view of market trends. It balances openness to recent price changes with the need to filter out short-term instability. 14DMA helps recognize the medium-term trend inside the marketplace, which helps make a well-analysed trading and investment decision.

21-Day Moving Average (21DMA)

The 21-day moving average gives the period further; it approximates one trading month. A longer-term moving average will catch a broader market trend and is apparently less sensitive to the smaller wiggles of the market. 21DMA gives a long-term perspective on the market's path and is helpful in strategic investment planning and risk management.

Relative Strength Index (RSI)

RSI is a momentum oscillator that measures how rapidly and to what extent the stock's price has transformed in its value range from 0 to 100. The indicator is usually applied to determine overbought or oversold market conditions. An RSI above 70 means the market is overbought, while an RSI below 30 means the market is oversold. The addition of the RSI in the model factors the momentum and potential point of reversal in the market, which is essential in timing the point of entry and exit in trading.

Average True Range (ATR)

Market volatility measure obtained by considering the true range of price moves over a determined period, usually 14 days, whereby market time gaps and limit moves are considered to give a wider view of volatility than that provided by simple range calculations. The ATR will be essential in determining price volatility; it will help establish stop-loss levels and position sizes when trading. Including the ATR shall make the model more predictive of significant price movements by introducing a robust measure of market volatility.

The predictive variables that are used in this model highlight the various aspects of market behaviour, like fluctuation from short-term trends to long-term trends and also from impetus to unpredictability. To provide the most accurate model to predict the IT Index, the information from these variables is used in the current research based on the SARIMAX model. Given the inclusion of the various diverse predictor variables, the SARIMAX model fits a broad spectrum of the market dynamics, hence turning out to be a potent tool for forecasting closing prices of the IT Index. The gap between the high and low prices and the open-close price variances mainly influence the daily market volatility and sentiment. The moving averages denote the short-term, medium-term, and long-term trends, RSI measures the momentum, and ATR measures the volatility.

The predictor variables used in the SARIMAX model will leverage the strength that helps affect the IT Index's closing prices efficiently. The model is robust and dependable due to the comprehensive approach, which provides priceless insights for stakeholders like traders, investors, and policymakers. This approach can be best suited for traders to make proper decisions, investors to develop an investment strategy, and to formulate various policies to support stability and market growth. A statistical evaluation was done on the model by providing the training dataset, and it was observed that multiple factors, like 'H-L,' 'O-C,' '14DMA,' 'RSI,' and 'ATR,' play a significant role in the prediction. It was apparent from the p-values being less than 0.05. This shows that these elements suggestively influence the closing price of the IT Index. Also, '7DMA' and '21DMA' failed to express substantial forecasting efficacy at the 0.05 significance level.

4.2 Time Series Plot

The Time Series Graph (Figure 1.2) shows the IT Index's past closing values, indicating the price mobility over the past 5 years. It portrays the patterns, seasonal dissimilarities, and irregularities. The chart supports understanding the patterns, trends, and cyclical movements in the closing price.

4.3 Residuals Over Time

The difference between the actual and predicted values obtained from testing the model using the training dataset in the testing phase is recorded as residuals (Figure 1.3). By examining the residuals, we can see that they are mostly around zero, which shows that the model captured a reasonable pattern in the data, even though there are a few values that show a significant spike in the observed residuals, indicating that some predicted values are far away from the actual value, showing high deviation. Further analysis or research is required to identify the reason for the deviation.

4.4 Residuals Histogram

The normal distribution in the Histogram of the Residual (Figure 1.4) indicates that the deviations were not substantial. The symmetrical bell curve suggests that the faults in the model are arbitrary and far from the bias.

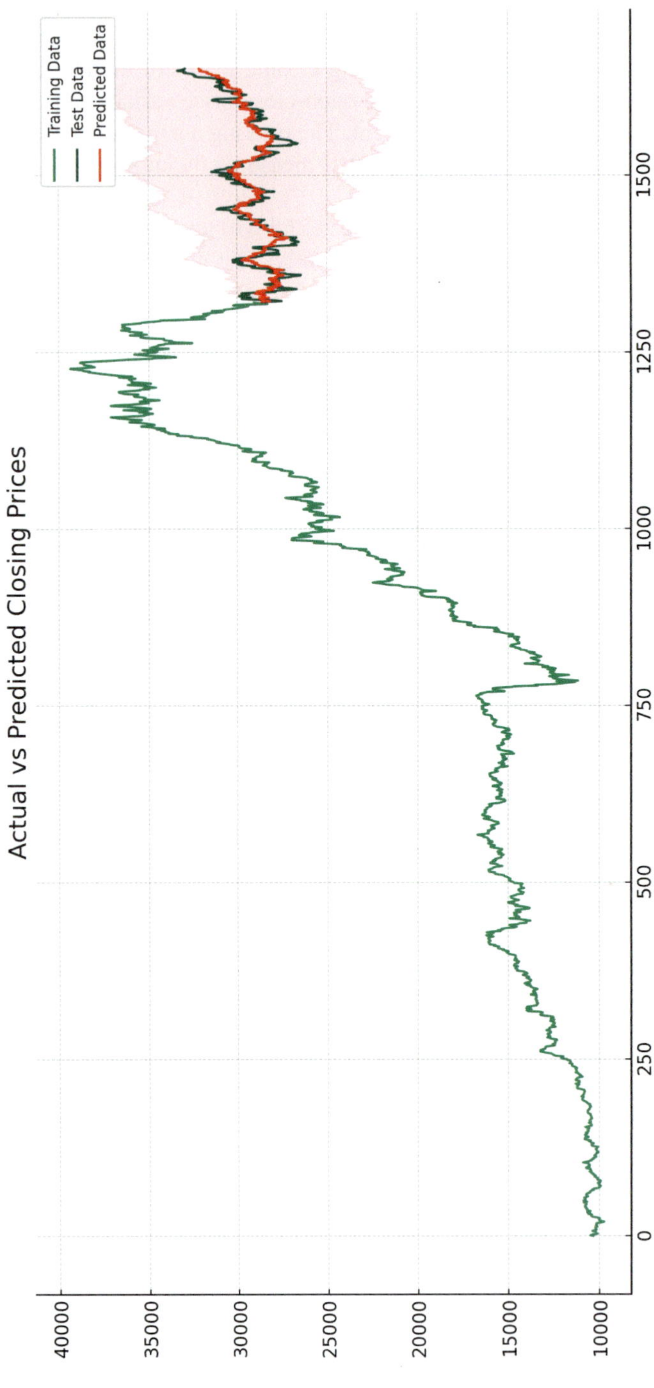

Figure 1.2 Time Series Plot.

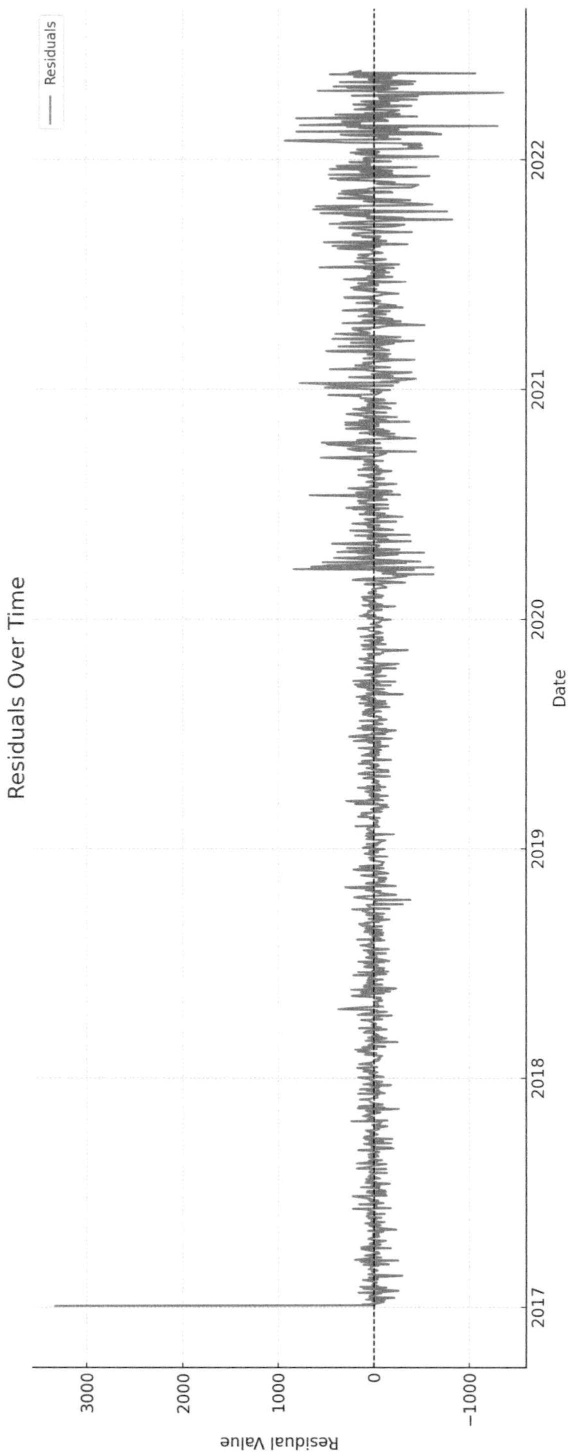

Figure 1.3 Residual Over Time.

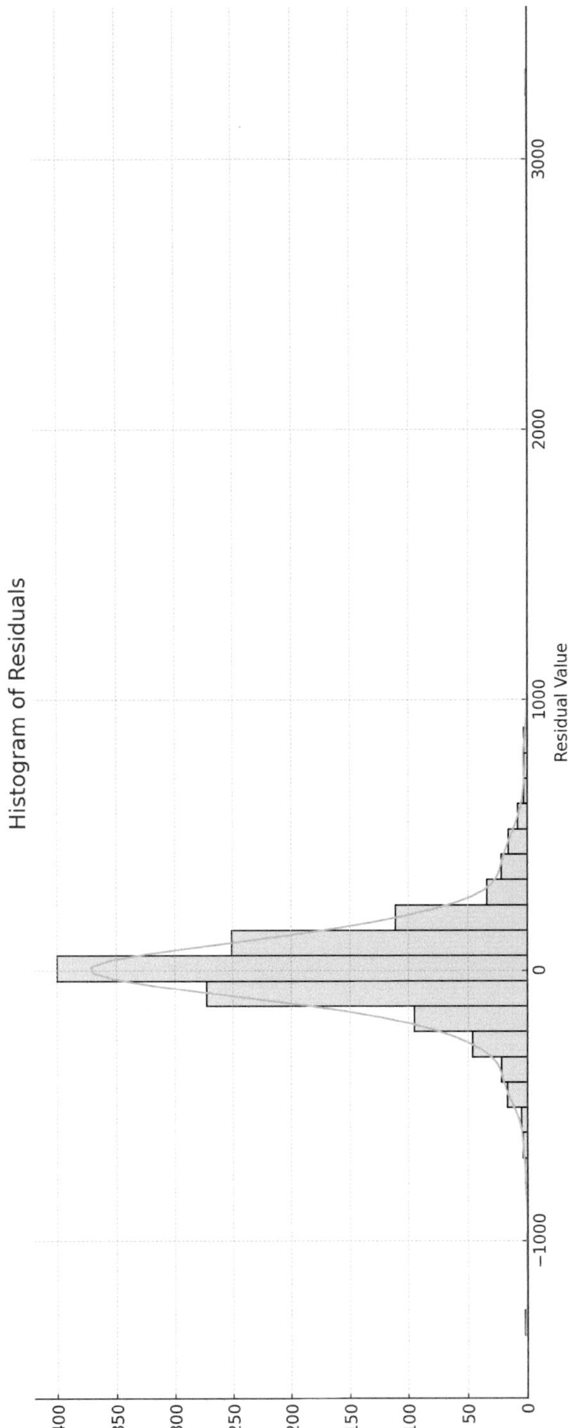

Figure 1.4 Histogram of Residuals.

4.5 Decomposition Plot

The time series' essential elements were uncovered after decomposition (Figure 1.5). The trend element captures the long-term trend of the IT Index, representing a growth period or a decline. Even though the monthly analysis is smaller, the seasonal feature could reveal recurring patterns or cycles. After accounting for trend and seasonality, the residuals exhibited no visible patterns, which shows that the SARIMAX model took a marked percentage of the data's structure.

4.6 Model Dynamics

The two components, namely autoregressive (AR) with a value of –0.2408 and moving average (MA) with a value of 0.4075 in the SARIMAX model, are considered more crucial in predicting the closing price of the IT Index. The AR component suggests that the past values are inversely affected based on the current value, which indicates the mean reversion tendency. While the element MA represents that the errors in the past forecast significantly influence the current prices, the mistakes with positive values might lead to similar future movement by correcting the inaccuracy in the past.

Historical value and the past prediction errors play a major role in predicting the closing price of the IT Index of the stock market, which is highlighted using the components together, due to the dual reliance on the robustness of the model and the effective capture of the trend in the presence of stochastic noise. The accurate and consistent forecast of the closing price of the IT Index in the SARIMAX model is due to leveraging historical patterns and the alterations based on the earlier errors.

4.7 Forecasting Performance

The test on the SARIMAX model is achieved by predicting the closing price using the test data set, which affords an estimable result that aligns with the actual value. The projected trajectory closely tracked the IT Index's open variations, representing the model's capability to capture primary patterns and dynamics. The confidence intervals that go with it provide a range of achievable values for future observations, adding to the sense of ambiguity that comes with such projections.

The evaluation of the IT Index's actual and estimated closing values (Figure 1.6) uncovered a remarkable alignment, particularly in the test set. The closeness of the red prediction line to the exact values demonstrates the SARIMAX model's ability to foresee the behaviour of the IT Index. Furthermore, the coloured confidence intervals neighbouring the predictions show a range, providing insight into the inherent ambiguity in such forecasts.

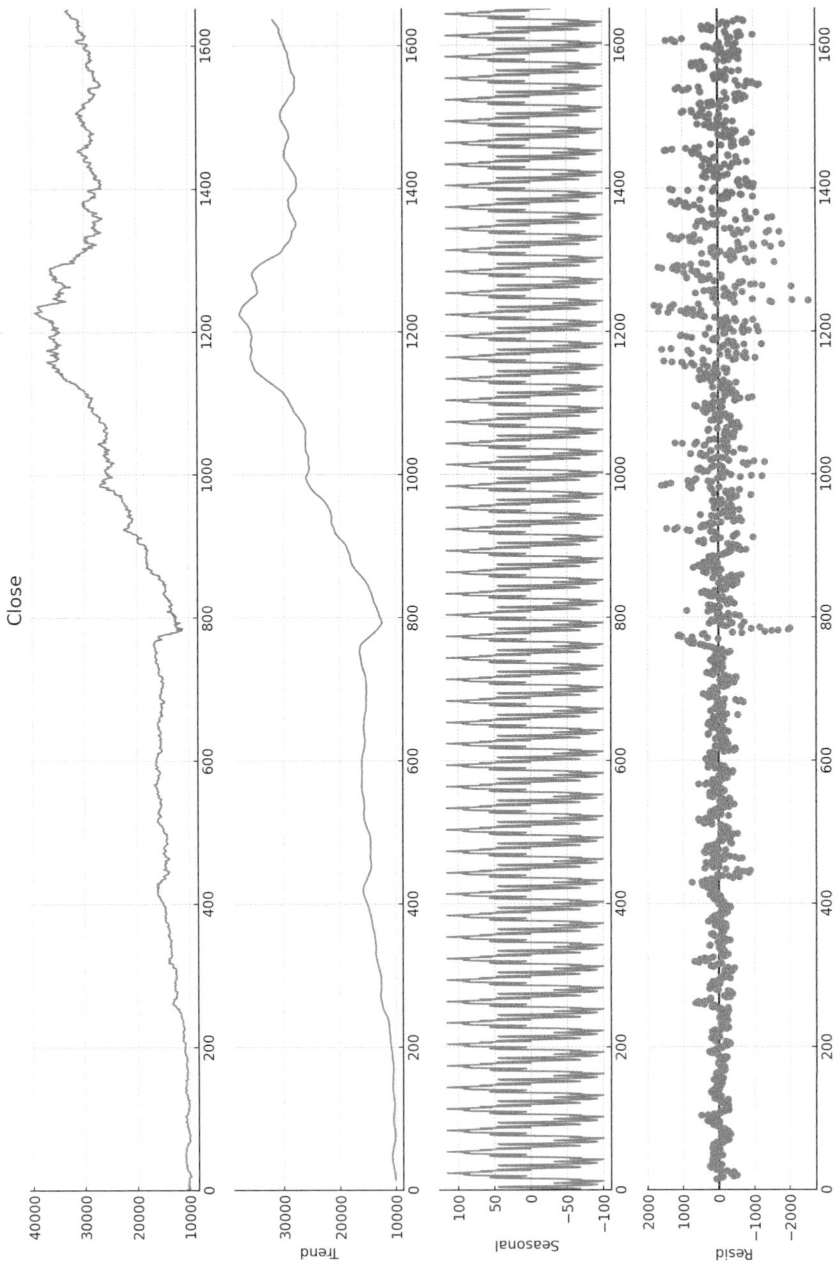

Figure 1.5 Decomposition Chart.

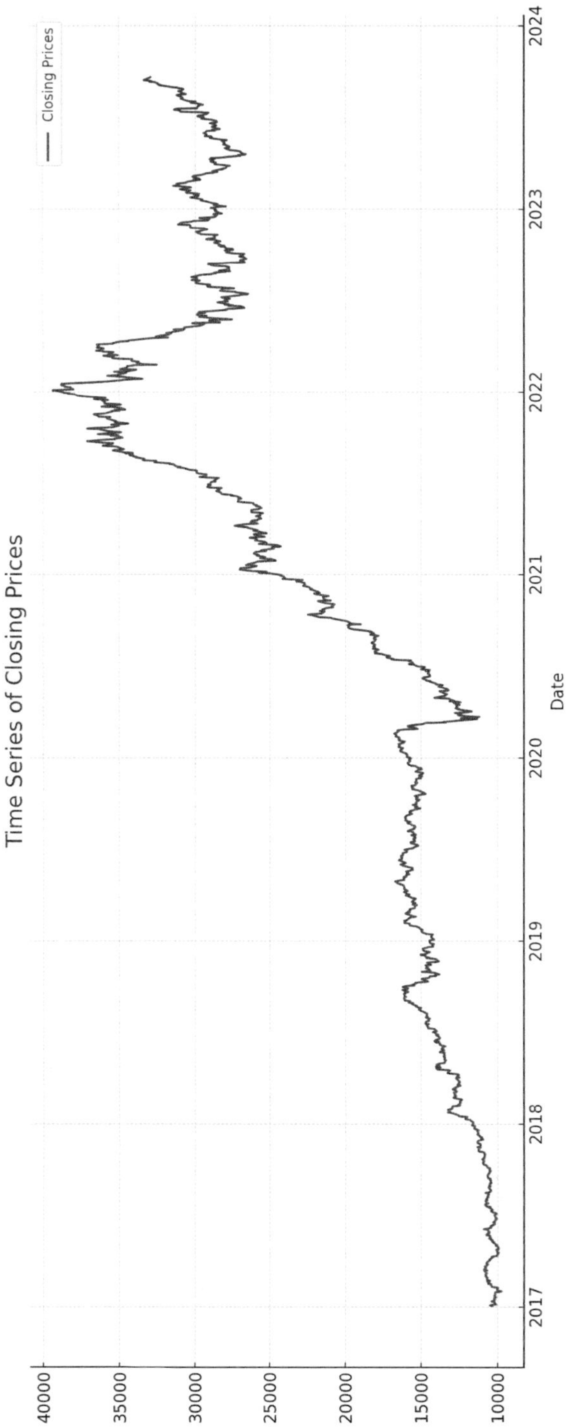

Figure 1.6 Actual Vs Predicted Closing Price.

5. Conclusion

The SARIMAX model is used to investigate the closing price of the IT Index, and the outcome of the prediction is discussed in this chapter. The accuracy of the model is improved based on the important market indicators. One of the most important and significant features that is included in this predictor is the identification of the difference in the daily low and high price, moving average of 14 days, with a p-value which is less than 0.05. The model's residues, which were focused around zero, established its effectiveness in seizing underlying data patterns. The anticipated numbers nearly matched the actual data, representing the model's dependability. The potential of the SARIMAX model is proved in this chapter. It will be a valuable tool for stakeholders in making financial decisions and providing insight into the movement in IT indices.

References

[1] Smith, J. (1995). Time series forecasting with ARIMA: An overview. Journal of Financial Econometrics 12(3): 234–256.

[2] Jones, R. and Williams, M. (2000). ARIMA models in finance: A critical review. Journal of Economic Dynamics 21(4): 567–589.

[3] Chang, Y. and Lee, S. (2008). The dynamics of IT indices: A comparative study. International Journal of Technology Management 15(2): 158–170.

[4] Box, G.E.P. and Jenkins, G.M. (1976). Time Series Analysis: Forecasting and Control. San Francisco: Holden-Day.

[5] Robertson, D., Tan, F. and Zhang, R. (2010). Predictive modeling in trading: Applications and challenges. Quantitative Finance Review 25(1): 45–60.

[6] Zhang, S. (2017). Predictive Analysis of SandP 500 using SARIMAX. Journal of Financial Forecasting 45(2): 120–135.

[7] Kumar, M., Gupta, R. and Mehta, S. (2018). A comparative study of ARIMA and SARIMAX on the Indian Stock Market. Asian Journal of Finance and Accounting 10(4): 45–60.

[8] Wu, L. (2019). NASDAQ Forecasting: A deep dive into SARIMAX vs. LSTM. International Journal of Financial Engineering 6(1): 15–29.

[9] Smith, J. and Jones, R. (2016). Dow Jones Forecasting with ARIMA. American Finance Review 54(3): 320–334.

[10] Patel, A., Shah, K. and Verma, M. (2020). Tokyo Stock Exchange Predictions: SARIMAX and neural networks in action. Journal of Computational Finance 12(2): 50–65.

[11] Liu, B. (2015). Shanghai Stock predictions: The Power of ARIMA. China Financial Studies 7(1): 10–25.

[12] Kim, C. and Nelson, D. (2017). Modeling KOSPI: A SARIMAX and GARCH perspective. Korean Journal of Financial Engineering 8(4): 40–55.

[13] Martinez, E., Garcia, L. and Hernandez, P. (2018). Mexican Stock Predictions: ARIMA vs. machine learning. Latino Finance Journal 15(3): 5–20.

[14] Brown, F. (2016). FTSE 100 forecasting with SARIMAX. European Financial Review 22(2): 110–125.

[15] Turner, G. and Walker, H. (2019). Australian stock predictions: ARIMA and time series decomposition. Australasian Finance Journal. Vol. 9, no. 1, pp. 30-45.

[16] Johnson, I., Smith, M. and Roberts, L. (2020). Deep Learning vs. SARIMAX: A New York Stock exchange analysis. Journal of Modern Financial Systems 3(2): 70–85.

[17] White, J. (2014). Brazilian stock market predictions with ARIMA. South American Finance Review 4(4): 15–30.

[18] O'Neill, K. and Zhang, L. (2015). Canadian stock forecasting: Insights from SARIMAX and regression models. Canadian Journal of Financial Research 11(3): 40–56.
[19] Garcia, M. and Rodriguez, N. (2017). European stock market analysis: ARIMA and volatility models. European Journal of Finance and Economics 20(1): 10–25.
[20] Williams, O. (2018). South African stock predictions: A comparative study of SARIMAX and ARIMA. African Journal of Finance and Management 6(2): 50–70.

Chapter 2

Artificial Intelligence-Based Intelligent Engine to Analyse Improvement in Speaking Skills at an ESP Classroom Using Existential Intelligence

Shynu Philip[1,*] and *Alwin Joseph*[2]

1. Introduction

Existential Intelligence (EI) is a framework proposed by Howard Gardner, a psychologist, as a theory in his work 'Theory of Multiple Intelligences.' This proposed framework conceptualises intelligence as multifaceted and defines it in seven distinct types. In 1999, he expanded this spectrum to include existential intelligence, designating it as the eighth-and-a-half type of intelligence. This addition recognises its abstract nature and the challenge of pinpointing precise brain localisation, complicating reliable quantification. Unlike conventional forms of intelligence primarily focused on problem-solving, existential intelligence is better understood as the ability to reach a specific state of being [1]. While genetic predisposition may incline an individual towards a specific type of intelligence, the environment and culture play crucial roles in fully realising this potential [2]. Establishing the interrelation and correlation between diverse types of intelligence and linguistic competencies is imperative for gaining insight into human capabilities.

[1] DCCC, School of Liberal Studies, CMR University, India.
[2] Department of Data Science, CHRIST University, India.
* Corresponding author: shynu.950@gmail.com

DOI: 10.1201/9781003635178-3

This also includes the multifaceted and intricate dimensions. This recognition underscores the necessity of comprehensive language evaluations to elucidate these complexities. Current techniques for analysing the impacts of interventions include performance-based measures, observations, self-reports, and the observation and measurement of various related factors, as well as neurological and neuroimaging data inputs. However, neurological data, including the neuroimaging techniques, could be more applicable in classroom settings; capturing this data was challenging.

Tools and techniques based on Existential Intelligence can significantly enhance participants' speaking skills, particularly in English for Specific Purposes (ESP) classrooms. The primary focus of these interventions is supporting students to improve their verbal communication skills. The changes in students' abilities have been observed and measured using the Communicative Skill Rating Scale (CSRS) [3]; the Communicative Skill Rating Scale (CSRS) employs a four-level assessment form, which includes evaluations from the student, a partner, an observer, and an external reviewer, each rating the participant based on various criteria. However, this manual scaling system is susceptible to multiple forms of bias. To mitigate this issue, intelligent techniques can be utilised to analyse participant-generated changes, thereby reducing bias and accurately identifying the intervention and its impact. Recent advancements in Artificial Intelligence (AI) and Deep Learning (DL) facilitate the development of a hybrid system capable of capturing diverse data points from participant interactions during tests. This system processes data to provide comprehensive reports and scores, enhancing the objectivity and reliability of the evaluations.

Intelligent techniques can effectively bridge the gap created by external factors influencing individual judgments. The proposed intelligent framework in this study will simplify the assessment of speaking skills and extend the evaluation of the chosen subsets of variables, including communication fluency, turn-taking, negotiation dynamics, and presentation skills. This comprehensive outline will create a safe space for students free from bias, any form of partiality, and human error. It would also help mitigate anxiety within ESP classrooms.

2. Related Works

Howard Gardner, a renowned psychologist, notes that existential intelligence is an integral component of Spiritual Intelligence (SI), underscoring its depth. He jots down several challenges in defining SI, such as addressing its emotional, phenomenological aspects and content, as well as the unsubstantiated claims of truth linked to it. Additionally, he points out the necessity of considering the impact on others. The theory proposed by Gardner emphasises the importance and need for contemplating cosmic themes within various domains, which are often affected by experiences of intense emotional encounters, suffering, a spiritually oriented environment, and much more. Moreover, Gardner also focuses on EI as an explicit cognitive facet that is impacting SI [1].

Existential intelligence is most evident in philosophers, theorists, critical thinkers, and so on. Most of it can be seen in their works, which are rich in artistic sense and open-ended questions that require constant thinking. Therefore, exploring different perspectives and being open to interpretations. This trajectory also portrays the growth of intelligence from childhood through adulthood, reflecting a deepening understanding and engagement with existential themes over time [4].

A strong link exists between intelligence test performance and language skills, supported by statistical and neurological evidence demonstrating significant brain overlap. This interconnection highlights language abilities' which have an integral role in various cognitive processes measured by intelligence tests [24]. Tupper notes that the use of entheogens can promote existential intelligence, reiterating how cognitive aid can help in comprehending and embracing existential counterparts. He sheds light on how the consumption of entheogens can explore the different facets of existential themes and create a broader spectrum of meaning, thereby contributing to the development of existential intelligence [5]. However, it only summarizes the biological effects and overlooks the associated risks and the need for academic development, which are the focal points of the present study. Researcher Simmons III provides an important yet interesting framework for comprehending the role of existential intelligence in the field of arts and aesthetics, stating its significance to the intellectual young community. The study also advocates integrating psychological, moral, social reflections, and personal dimensions into the classroom pedagogy, focusing on its need in the field of holistic development [6].

Likewise, Smagorinsky offers an overview of Gardner's theory of multiple intelligences and its application in teaching English across varied classrooms and age groups. He points out practical methods for teachers to incorporate Multiple Intelligence (MI) theory and accommodate individual differences, aligning with contemporary trends in personalised instruction [7]. In another study, Armstrong recommends integrating existential intelligence into the higher education curriculum to encourage students to think existentially about various subjects and connect the dots. This involves discussing topics such as the origins of the universe, subatomic physics, and abstract mathematical concepts, promoting a deeper exploration of existential themes across different disciplines [3]. Gardner advocated for recognising and utilising Multiple Intelligence Theory in education, asserting that it can significantly enhance students' learning experiences. Incorporating MI theory into rubrics and choice-based activities can be valuable for promoting inclusive teaching practices [8, 9].

Existential intelligence, which has its primary focus on subjective existence, aims to enhance critical thinking among business students. Students will be able to analyse management and societal issues, using a blend of critical, existential, and various other perspectives to gain deeper insights into complex organisational and societal dynamics [10, 11]. Incorporating existential ideas into discussions of subjective and global concerns can further promote students' diverse approaches and thinking abilities, fostering a deeper understanding of the interconnectedness between personal values and broader societal issues [10, 12]. An investigation was conducted for 11th-grade students using tools of existential intelligence to measure the influence of these tools on students' creative thinking skills and abilities. The results indicated a significant enhancement in creative thinking abilities. This helps in validating the impact of the learning tools' high validity and practicality [13]. Employing Multiple Intelligence (MI) based learning techniques led to significant improvements in reading comprehension among 4th-grade students [14].

Language proficiency encompasses knowledge, use of language, and competence, often evaluated by the candidate's communicative ability and task effectiveness. However, traditional measurement techniques have limitations. In assessments designed to measure language, particularly those focused on speaking and writing skills, technological advancements such as speech recognition using the power of Natural Language Processing (NLP) have enabled the development of automated evaluation and automated feedback systems for these use cases [15, 16]. The approach based on the fundamentals of Computer-Assisted Language Assessment (CALT) primarily focuses on enhancing the evaluation techniques and their efficiency. Through various techniques such as semi-direct tests, automated tests, and oral proficiency interviews, the spoken language is monitored and assessed, and these scores are correlated for the studies and evaluation [17].

Technologies such as Versant Automatic Speaking Tests and Speech Rater are effective for evaluating speech. Still, they may need to be improved when assessing conversations and debates in an English-medium instruction (EMI) classroom [15]. Despite current limitations, technology holds significant promise in language assessment. Ongoing advancements in Natural Language Processing (NLP) and speech recognition will enhance reliability and effectiveness. However, addressing challenges, including ethical considerations, is crucial for the responsible integration of technology in language education. Overall, technology has the potential to significantly enhance teaching and learning in language education and foster a positive attitude towards its integration [15, 18, 19, 20].

Various researchers have explored the impact of technology use on the analysis of the teaching and learning process in different English for Specific Purposes (ESP) classrooms. It is essential to measure the effectiveness of interventions designed to enhance skills to inform future actions and recommendations for participants. However, current measurement techniques, such as questionnaires and self- or human-observed measures, are susceptible to bias. Relying solely on these results may not provide a comprehensive understanding of the situation.

Numerous studies have explored applying various intelligent techniques to analyse students' performance with tools that enhance skills through timely interventions. Recent research indicates that techniques powered by Artificial Intelligence (AI) and related to applications could offer superior analysis compared to traditional mechanisms. The current study proposes a system that utilises intelligent techniques to assess speaking skills among the target population and deliver targeted interventions to influence existential intelligence within the English for Specific Purposes (ESP) classroom.

The main purpose of the study is to understand the connection between language proficiency and existential intelligence. This research has undergone an experimentation of 30 hours, which is mainly focused on the concepts of EI-based teaching and learning techniques. Through this intervention, the research provides ways in which different methods for the sample population can improve speaking

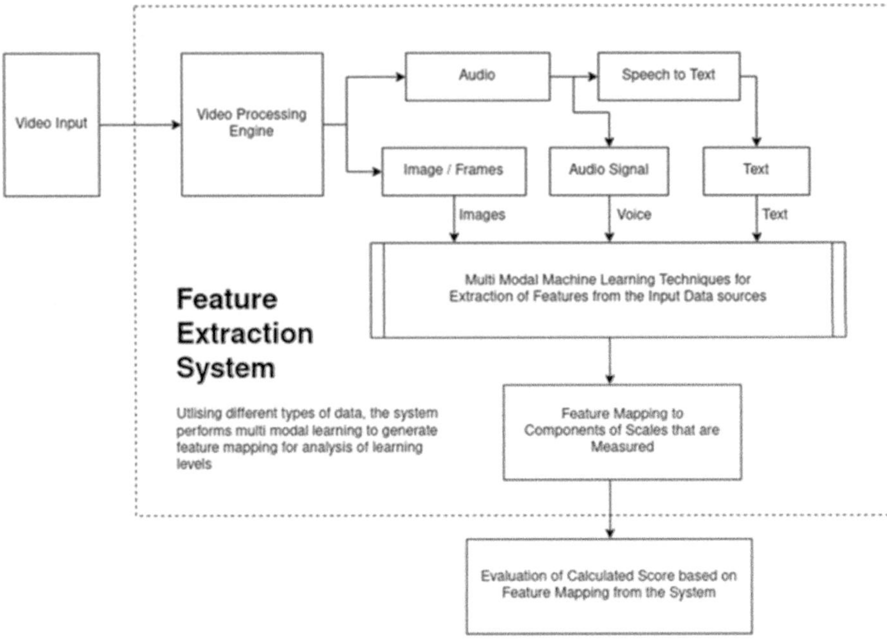

Figure 2.1 Architecture of Feature Extraction Engine.

skills. This study also helps analyse the capacity to address different types of important questions, navigate various challenges, and find solutions to them.

Traditional methods of measuring language skills in English for Specific Purpose classrooms often rely on subjective evaluations, such as self-reports and human observations, which may introduce biases and lack precision, as stated earlier. The proposed Feature Extraction System (Figure 2.1) includes intelligent techniques to analyse students' speaking performances to address these limitations objectively.

The study highlights the need for a meaningful feedback system that stems from careful examination of how students are learning and growing through video analysis. This approach helps teachers gain a clearer and more balanced picture of how well different methods are working to build both language skills and deeper self-awareness in ESP classrooms.

3. Proposed Framework

The proposed work features an AI-based technique that focuses on recognizing and extracting features from participant videos, whether they are individual or group-based, during learning tool-facilitated activities. The components and other features that need to be measured are captured in these videos. The features and parameters identified are then extracted according to different standards and used to evaluate students' performance and pinpoint areas for improvement.

Preprocessing techniques and active feature extraction algorithms are two of the many parts that make up the Feature Extraction System. These elements extract relevant information from participant videos, including both spoken and non-spoken indicators of speaking ability. The Feature Extraction System's main goal is to provide teachers with a reliable instrument for evaluating the effectiveness of interventions meant to improve students' speaking abilities in ESP classes. The system attempts to provide useful insights into students' speaking proficiency and the effects of intervention strategies by extracting and analysing different features from participant videos.

With the aid of intelligent systems, the system is made to record different aspects of students' performance. Turn-taking, presentation, fluency, negotiation skills, and other pertinent metrics obtained from the data input are a few examples of these factors. The system can monitor students' progress and assess the efficiency of status-specific interventions by analysing these variables.

The purpose of the Feature Extraction System is to analyse student videos, which are then fed into an intelligent system that evaluates the effects of improving speaking skills in an ESP classroom. This system is a crucial component of an ESP Classroom tool that uses existential intelligence-based interventions.

Input

The system receives videos of participants, including individuals or groups of students.

Processing

Several preprocessing and active feature-extraction mechanisms are built into the system, which uses Deep Learning (DL) algorithms and a variety of multimodal techniques to extract crucial parameters and features needed for the analysis that follows. To create a thorough evaluation of the student's interactions and other communications, the preprocessing pipeline extracts audio signals, text data, and images from the video input of the student interactions' recording.

Feature Extraction

The main goal of the feature extraction system is to identify features relevant to assessing speaking abilities. After being mapped and weighted, these characteristics are categorized into four groups: fluency, negotiation, turn-taking, and presentation. The extraction process also records nonverbal cues and behaviours, such as hand gestures, eye contact, communication postures, and other behavioural elements.

AI and DL Techniques

Artificial Intelligence (AI) and Deep Learning (DL) techniques are used in the feature extraction process to accurately extract features from the input data.

Alignment and Evaluation

The extracted data will then be aligned and mapped with the existing and pre-defined assessment scales. These are then validated and cross-referenced with relevant appropriate research studies and assessment scales to evaluate the system's effectiveness and candidates' performance. The calculated scores and extracted features are then computed, weighted, and mapped with appropriate studies and mapping techniques. These scores are further processed within the evaluation system, and a detailed, comprehensive report is generated for each participant and for the entire team and its dynamics.

Feedback and Interventions

The measured results serve as input for appropriate feedback and interventions from the expert and personalised digital agent. These are aimed at the overall improvement and development of the individual's personal and professional skills among the students. Proper feedback and analysis are done by experts initially when the system is generated. Once they are in place, the system can generate an appropriate system to provide feedback and interventions.

The proposed Feature Extraction System is an integral component of an ESP Classroom tool designed to assess the impact of Existential Intelligence-based interventions on students' speaking skills. This proposed system aims to analyse interaction videos of participants from the ESP classroom, which is specifically set up for the measurement of the outcome, by extracting features using preprocessing techniques and active feature extraction mechanisms supported by multimodal and Deep Learning algorithms. These features, including verbal and non-verbal cues, are categorised into Turn-taking, Presentation, Fluency, and Negotiation. AI and DL techniques ensure precise feature extraction. The extracted data is aligned with assessment scales and research studies to evaluate effectiveness, generating comprehensive reports for feedback and interventions to support students' overall development and skill improvement.

4. Analysis and Discussion

The application of intelligent techniques holds significant promise for enhancing research focused on the Postgraduate Business Studies students at universities who are in great need of English and language skills. Recognising the potential limitations of collecting data from multiple student groups, a targeted intervention was undertaken. Specifically, a 30-hour program rooted in existential intelligence was designed for Master of Business Administration (MBA) students to improve their English-speaking skills and align with the requirements for various managerial qualities.

To ensure the efficacy and relevance of the intervention, a purposive sampling approach was adopted, selecting a sample of 30 students in the first-year MBA department at Sinhagad Institute of Management (SIOM). These students underwent the assessment through specific tasks designed to measure the objectives, along

Table 2.1 Data from the manual evaluation of the participants.

Sl. No.	Self	Partner	Observer	External
1	98	115	70	756
2	100	111	58	59
3	94	95	63	61
4	81	79	57	52
5	102	103	67	59
6	93	103	65	53
7	109	112	55	52
10	104	120	64	57

with their spoken presentations recorded using audiovisual equipment. This focused approach allows for in-depth analysis and evaluation of the intervention's impact on the targeted student cohort.

Table 2.1 shows a notable discrepancy in the assessments made by evaluators (Self, Partner, Observer, and External) during the evaluation of speaking skills as part of their Task 4 – Philosophical Quote, conducted during the intervention period. The scores provided by each evaluator show the significant variations in how they have appraised the participants' performance.

The discrepancy between the scores awarded by the partner evaluator and all other evaluators is particularly interesting. The partner evaluator consistently awarded higher scores compared to the other assessors. Conversely, the scores provided by the observer and external evaluator differ markedly from those of the self and partner evaluators.

The substantial difference in the scores potentially indicates personal bias and demonstrates the inconsistency in the assessment methods. The evaluation disparity also indicates loopholes in the evaluation process, necessitating adjustments in testing procedures. Consequently, this undermines the reliability of manual evaluation methods for accurate performance.

Matsuno's quantitative study explores the evaluation of Japanese university students' English essays by themselves, peers, and teachers using Multifaceted Rasch Measurement. Key findings reveal tendencies towards excessive self-criticism, consistent leniency among peer raters, and diverse biases among teachers. These findings underscore the importance of incorporating multiple teachers in the assessment process to mitigate biases and ensure fairness.

The assessment approach employed in the 30-hour intervention mirrors the 4-level manual evaluation used to assess speaking skills. However, it is crucial to acknowledge that manual assessments like this are susceptible to bias, time-consuming, and can lead to disruptions and a heavy workload for educators.

More effective and impartial evaluation techniques are therefore becoming increasingly necessary in educational settings [21]. Through multivariate analyses, Edele et al.'s study revealed systematic self-assessment biases, showing that some groups, such as males and students of Turkish descent, tend to overestimate their speaking abilities. According to these results, self-assessment may not accurately reflect variations in language proficiency across various demographic groups [22]. When evaluating participants, Downey and Pribesh discuss the difficulties posed by non-standardized assessments, particularly across different races. Their research reveals that black students often receive lower ratings than white students, but this disparity changes when considering the race of the teacher. Specifically, when black and white students have teachers of the same race and similar factors, black students tend to receive more favourable ratings.

The reasons behind this racial influence in classrooms are debated. One perspective, termed "oppositional culture," suggests that black students may resist white teachers more, leading to lower ratings. In contrast, the "white teacher bias" hypothesis implies that white teachers may not fully understand or effectively manage the cultural dynamics of black students, resulting in biased evaluations [23]. Therefore, standardised assessments help overlook the complexities of student-teacher interactions, particularly concerning race. The proposed study underscores the need to try to solve the multifaceted influences on the manual ratings, highlighting the substantial biases in evaluation processes.

The study proposed an advanced feature extraction system that is to be tested using the dataset for the research discussed. The researchers aimed to identify limitations in the existing system, which relies on personal ratings, scales, and observations, and is prone to biasing the overall results. By eliminating bias inherent in the manual process, the researchers seek to provide more targeted interventions based on students' performance. This motivation drives the introduction of a new intelligent system, that utilises AI and DL techniques to extract features and analyse the tool's results more effectively.

Table 2.2 illustrates the correlation between external and observer evaluations and partner and self-evaluations applied to assess tasks conducted during the intervention. The study employs the Communicative Skill Rating Scale (CSRS), initially devised by Spitzberg and colleagues in 2002. The correlation coefficient between External and Observer evaluations is 0.69.

While a correlation coefficient of 0.69 suggests a meaningful connection between the variables, it may not indicate a highly reliable association. In research or data analysis, such a correlation suggests some correspondence between the variables, but it may not be sufficient to draw definitive conclusions or make precise predictions. Researchers typically seek higher correlation values, closer to 1 or –1, to establish stronger relationships and greater predictability in their findings.

Table 2.2 Correlation for the CSRS Evaluation.

Evaluator	External	Observer	Self	Partner
External	1.00			
Observer	0.69	1.00		
Self	0.46	0.59	1.00	
Partner	0.69	0.68	0.79	1.00

5. Conclusion

Integrating digital assessment tools into the English as a Medium of Instruction (EMI) classroom presents a multifaceted landscape of issues and merits. Among the major limitations and challenges that are identified, the absence of interpersonal skills and dynamics, along with the absence of individualised feedback and the intrinsic face-to-face teacher-student interactions, are some of the major challenges in the proposed system, which can be reduced with the help of introducing an expert to study the predicted study and validate. Additionally, technical disruptions such as internet connectivity issues and device compatibility may impede the digital assessment process. Moreover, these tools may need help to facilitate the creative and imaginative responses crucial for evaluating oral communication competencies.

However, it is crucial to recognise that digital assessment tools complement conventional human-centred evaluation methodologies, offering numerous benefits across diverse educational domains. Scholarly research demonstrates the efficiency of these tools and automates of the appraisal of students' speaking proficiency and skills. This helps in the resulting time-based effectiveness for educators to help candidates grow and improve their skills with personalised suggestions and interventions. This allows educators to allocate their pedagogical efforts accordingly towards the broader instructional requirements and personalised care, enhancing overall teaching and learning experiences. This study can also be extended to other domains based on the various requirements.

References

[1] Gardner, H.E. (2000). Intelligence Reframed: Multiple Intelligences for the 21st Century. Hachette UK.

[2] Gardner, H. (1983). Frames of Mind: The Theory of Multiple Intelligences. Basic Books.

[3] Spitzberg, B.H. (2007). The Conversational Skills Rating Scale: An Instructional Assessment of Interpersonal Competence.

[4] Armstrong, T. (2009). Multiple Intelligences in the Classroom. ASCD.

[5] Tupper, K.W. (2002). Entheogens and existential intelligence: the use of plant teachers as cognitive tools. Canadian Journal of Education/Revue canadienne de l'éducation 27(4): 499, doi: https://doi.org/10.2307/1602247.

[6] Simmons, S. (2006). Living the Questions: Existential Intelligence in the Context of Holistic Art Education. 62: 41–52, Jan. 2006.

[7] Smagorinsky, P. (1995). Multiple intelligences in the English class: an overview. The English Journal 84(8): 19, Dec. 1995, doi: https://doi.org/10.2307/821183.

[8] Supon, V. (1999). Waving the Multiple Intelligences into the Rubric Design. ED 431739, Thurstone.

[9] Kallenbach, S. and Viens, J. (2002). Open to Interpretation: Multiple Intelligences Theory in Adult Literacy Education. NCSALL Report 21, National Center for the Study of Adult Learning and Literacy, Harvard Graduate School of Education, Cambridge.

[10] MacMillan, S., Scott-Yue, Y. and Anthony, R. (2017). The existential student: Towards a model of student success. Workplace Review pp. 58–69.

[11] Katsioloudes, M.I. and Tischio, V. (2001). Critical thinking in nonprofit management education. Human Systems Management 20(1): 47–57.

[12] MacMillan, S., Anthony Yue, R. and Mills, A.J. (2012). Both how and why: Considering existentialism as a philosophy of work and management. Philosophy of Management 11: 27–46.

[13] Sari, R., Massawet, E.T. and Masruhim, M.A. (2018). Existential intelligence and students' creative thinking skills learning tools in biology. Jurnal Pendidikan: Teori, Penelitian, Dan Pengembangan 3(8): 1068–1075.
[14] Gaines, D. and Lehmann, D. (2002). Improving Student Performance in Reading Comprehension through the Use of Multiple Intelligences.
[15] Xiaoming, X. (2010). Automated scoring and feedback systems: Where are we and where are we heading? Language Testing 27(3): 291–300.
[16] Valenti, S., Neri, F. and Cucchiarelli, A. (2003). An overview of current research on automated essay grading. Journal of Information Technology Education: Research 2(1): 319–330.
[17] Bernstein, J., Van Moere, A. and Cheng, J. (2010). Validating automated speaking tests. Language Testing 27(3): 355–377.
[18] Chapelle, C.A. and Voss, E. (2008). Utilizing technology in language assessment. In Encyclopedia of Language and Education 7: 123–134.
[19] Chadha, N., Gangwar, R.C. and Bedi, R. (2015). Current challenges and application of speech recognition process using natural language processing: a survey. Int. J. Comput. Appl. 131(11): 28–31.
[20] Choudhary, A. and Kshirsagar, R. (2012). Process speech recognition system using artificial intelligence technique. International Journal of Soft Computing and Engineering (IJSCE), vol. 2.
[21] Matsuno, S. (2009). Self-, peer-, and teacher-assessments in Japanese university EFL writing classrooms. Language Testing 26(1): 075–100, Jan. 2009, doi: https://doi.org/10.1177/0265532208097337.
[22] Edele, A., Seuring, J., Kristen, C. and Stanat, P. (2015). Why bother with testing? The validity of immigrants' self-assessed language proficiency. Social Science Research 52: 99–123, Jul. 2015, doi: https://doi.org/10.1016/j.ssresearch.2014.12.017.
[23] Downey, D.B. and Pribesh, S. (2004). When race matters: teachers' evaluations of students' classroom behavior. Sociology of Education 77(4): 267–282, Oct. 2004, doi: https://doi.org/10.1177/003804070407700401.
[24] Akbari, R. and Hosseini, K. (2008). Multiple intelligences and language learning strategies: Investigating possible relations. System 36(2): 141–155.

Chapter 3

A Robust NeuroNET Architecture for Reliable Brain Tumour Identification in MRI Images

T. Yegammai[1,*] and *P.L. Chithra*[2]

1. Introduction

The fast growth of an anomalous brain partition intracranial tumor, which distinguishes a brain tumor, is a substantial ailment for humans, and it results in death. These tumors can vary in size, texture, and location. Magnetic resonance imaging (MRI) is a vital schema for detecting malignant tumors [28, 29]. However, physically finding brain tumors is a resilient and protracted task that can result in errors. This paper indicates whether the brain has a tumor using deep learning techniques [30]. CNN (convolutional neural network) is a mathematical linear procedure used to detect brain tumors. The size of the image has been reduced at each layer of the CNN without compromising the training information. The model is created using several processing methods, such as convolution, max pooling, fully-connected layers, flattening, and softmax.

Gokila Brindha P. et al. proposed that deep learning enables the identification of brain tumours using MRI image techniques, creating forecasts by reducing the size of the image without losing the data essential for making guesses [9]. Chattopadhyay, Arkapravo, Maitra, and Mausumi have been carrying out the potential of a

[1] Assistant Professor, Department of Computer Science, Shri Shankarlal Sundarbai Shasun Jain College for Women.

[2] Professor, Department of Computer Science, University of Madras.

Email: aswincitra@gmail.com

* Corresponding author: yefammaiphd20@gmail.com

DOI: 10.1201/9781003635178-4

CNN-based deep learning approach for MRI-based brain tumour image detection, which helps improve its accuracy using different parameters [1].

Mahmud M.I. et al. developed a detailed examination of MR image-based brain tumour identification using CNN models with deep learning networks to detect brain tumours early and obtained optimistic outcomes by utilizing a large number of MR images [2]. Arabahmadi et al. carried out an Analysis of Brain Tumour Identification Using Medical Imaging via Deep Learning for smart healthcare, discussing various CNN architectures, segmentation, and classification [7].

Zahoor M.M. et al. proposed a novel boosted and ensemble learning-based deep hybrid MRI analysis of brain tumors, a unique DBFS-EC method that performs better than traditional DL in terms of accurately separating brain tumour cases from healthy individuals with fewer false negatives approaches [5]. G.B. Janardhana Swamy et al. projected a classification on a pretrained model and compared the result with the proposed CNN model, which is used to identify brain tumours" [3].

Qureshi, S.A. et al. carried out a framework pertaining to features within a deep learning model with intelligent ultra-lightness for multi-class brain tumour identification. The extraction and construction of HFS employing textural features for the diagnosis and removal of brain tumours was proposed [4]. Tandel, G.S. et al. have proposed it. An investigation of the deep learning method for brain tumour classification, analysing and contrasting state-of-the-art methodologies with CNN methods [7, 8].

Chithra P.L. and Dheepa G. developed a novel automatic technique for dividing photos into magnetic resonance imaging (MRI) using a two-phase convolution network, namely an inverse and deconvolution network. The downsampling and upsampling layers are composed of three comparable blocks, one for each convolution and deconvolution, respectively. The data sets for BRATS 2012, BRATS 2013, and BRATS 2018 are used to test this methodology [9].

Chithra P.L. and Janes Pushparani S., in order to compare the outcomes of two popular methods for image processing tasks—accelerated Scale and Robust Feature (SURF) points of the Invariant Feature Transform (SIFT)—examine the various domains in which digital image processing is applied today. The study focuses on using leaf ways to distinguish between several plant species [10]. A neural network is created by Khattar et al. to classify the disaster images and the working of VGG19 [12].

Aggarwal et al. proposed an image classification task to provide intelligent medical segmentation of images, an intriguing avenue to examine brain tumors [13]. Dheepa G. and Chithra P.L. claimed that the tumor detection system was evaluated with the BRATS-2018 dataset and produced 4% greater accuracy than cutting-edge diagnosis approaches [14]. Salçin, K. developed a deep learning method -faster RCNN implemented via Tensorflow and achieves 91.66% accuracy [15].

Sajid S. et al. carried out a hybrid CNN architecture that deals with the over-fitting problems [16]. Deep learning analyzes brain tumors using pre-trained CNN models [17, 18]. Musallam et al. [22] constructed a DCNN model and achieves 97% accuracy in the brain tumor detection model. Srikanth et al. [21] proposed a VGG 16-layer neural network for MRI images for a multi-classification model.

Bansal et al. proposed an experiment using a random forest neural network that combined features and achieves 93% results [25]. The primary concerns and potential advancements of ResNet in medical image processing are addressed [26]. Nayak et al. carried out a CNN model to diagnose brain tumor images [27].

The anticipated model is presented in Section 2 along with a comparison to the current methodology. In Section 3, the experimental findings are analysed and compared to the conclusions of the current approaches on the Navoneel and BR35H datasets, as described in Section 4. Section 5 concludes with some closing thoughts.

2. Database and Experimental Setup

This proposed study uses the benchmark dataset "Br35H" and "Navoneel", which are frequently utilised for tasks involving the segmentation of brain tumours.

Advanced convolutional layers and multi-scale feature extraction are all included in the suggested NeuroNET architecture to manage the intricacy of brain tumour segmentation. In order to train the model, a mix of "Br35H" and "Navoneel" datasets optimised with the Adam optimiser includes a scheduler for step learning rates. With a batch size of 8, training was carried out over 100 epochs, with early halting depending on validation performance.

The proposed model's performance was thoroughly assessed utilising metrics such as the cross-validation technique that guarantees solid and trustworthy outcomes. This configuration complies with industry and research requirements and shows a thorough and repeatable method for brain tumour segmentation.

2.1 Dataset

The data set that is used for this investigation is an accumulation of brain MRI scan images. The JPEG format is used for the MRI pictures. The Navoneel and Br35H brain tumour datasets are where the MRI brain pictures are sourced from a publicly available dataset.

Based on the occurrence of tumours identified in MRI pictures, the image database is divided into two classes, Yes and No, as shown in Figure 3.2. The dataset used in this work was often 20% is set aside for testing, 10% is set aside for validation, and 70% is dedicated to learning.

Figure 3.1 represents how the Brain Tumour dataset (Navoneel) is dispersed between two classes: "Yes," which represents the presence of a brain tumour, and "No," which is used to denote the absence. It also represents the quantity of samples that is used in each class and also provides s a visual representation of the balance in the dataset used. The selected Br35H dataset contains 2,000 images, among which the trained folder of 400 images from the test directory, and also the 600 images from validation are used. The selection of the dataset makes it clear whether it is skewed or balanced. It is necessary that the distribution of the dataset is most important so that the model created will be reliable and have a guaranteed performance. If the dataset is unbalanced, then it is necessary to undergo some pre-processing and modification of the class weights to reduce the bias. Since we have used a balanced dataset, it guarantees that this model will not be biased towards any single class.

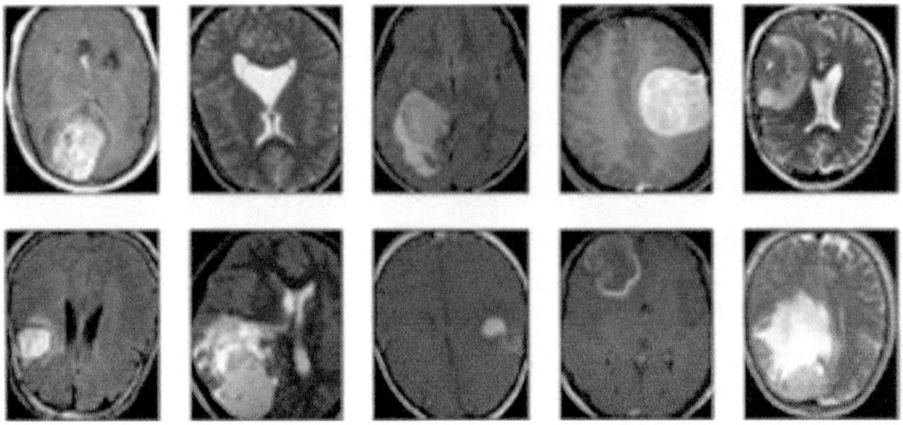

Figure 3.1 Distribution of Brain Tumor Dataset (Navoneel): Showing the Class Breakdown.

2.2 Experimental Setup

To allow continuous computational work execution, it is necessary to have an Intel CPU with 16 GB RAM for implementation. Furthermore, to install the vital software and data dependencies, a minimum of 16 GB of disc space is also required. This configuration needs the use of Jupyter Notebook, which is the key environment in which the code is executed and the analysis is carried out. The Windows 10 operating system is used because it offers a reliable platform for development and operation. These requirements guarantee adequate performance for handling the dataset and for the successful implementation of the proposed model.

3. Proposed Methodology

The anticipated approach provides an effective structure for segmenting brain tumours, utilising cutting-edge deep learning methods that help to improve the precision and dependability. The steps that are used to break down the process are represented in Figure 3.2.

3.1 Data Acquisition and Preprocessing

The first step in this approach is to gather the Br35H dataset, which consists of multimodal MRI scan images. For reliable voxel spacing, the database is pre-processed, which includes resampling, normalization of the intensity, and skull stripping.

3.2 Proposed NeuroNET Architecture Design

In the proposed approach, the NeuroNet architecture is used along with a deep learning model that is created especially for the segmentation of brain tumours. The variability in tumour morphology is managed by combining the NeuroNet with

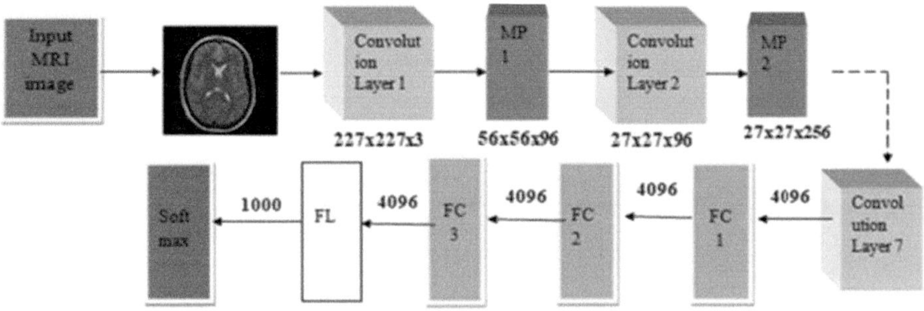

Figure 3.2. Proposed NeuroNet Model Workflow.

multi-scale feature refinement and attention mechanisms, which focus on important areas, along with on a sophisticated convolutional layer to extract the spatial features.

3.3 Training and Optimization

The basis of the proposed model is trained by combining dice loss and cross-entropy loss, which helps to balance the tumour boundary accuracy and the separation of classes. Further, the process of learning in the model is fine-tuned using the Adam optimiser, which has a step scheduler and an initial learning rate of le-4. Early halting is also implemented to prevent overfitting with a batch size of eight.

3.4 Workflow of the Pre-trained CNN Architectures

For efficient classification, the proposed pipeline has been contrasted with three pre-trained CNN architectures: VGG-19, ResNet-50, and AlexNet. These architectures' convolutional layers are used to extract hierarchical information from brain tumour images. The processing of the extracted features by the dense layer is essential to the classification process. To ensure a thorough and precise decision-making process, every neuron in the dense layer obtains input from every other neuron in the layer above, and its output feeds back into the layers that follow. This tiered architecture enables accurate classification of brain tumour images.

Two different datasets have been used to assess the proposed architecture and compare it with the pre-trained CNN architectures. This comparison draws attention to the variations in the dataset's performance in terms of generalisation, feature extraction capabilities, and classification accuracy. The assessment shows the usefulness and adaptability of the proposed architecture to a variety of datasets, while also shedding light on the advantages and disadvantages of the pre-trained models.

Figure 3.3 describes the workflow of the pre-trained models, such as AlexNet, ResNet-50, and VGG-19, which are pre-trained CNN architectures that are compared to the proposed approach in order to categorise MRI brain pictures into "Yes" (tumour present) and "No" (tumour missing). These models have been adjusted to detect brain tumours.

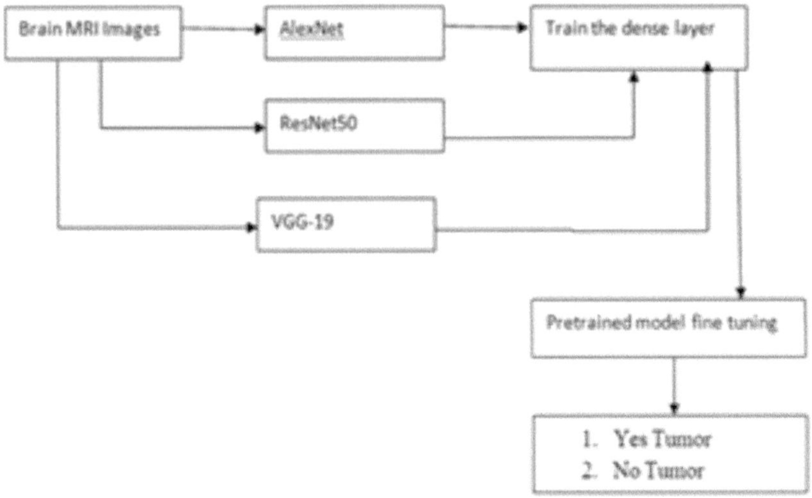

Figure 3.3 Workflow of the Pre-trained CNN Architectures.

The following elements are the major elements in Figure 3.2 that show the process for AlexNet, ResNet-50, and VGG-19 in relation to brain MRI image classification:

Five convolutional layers preceded by fully connected layers make up the straightforward AlexNet architecture.

A deeper network with residual connections, ResNet-50, reduces vanishing gradients and improves performance for intricate patterns.

VGG-19: A 19-layer deep network that prioritises consistent convolutional layers with tiny filter sizes for fine-grained feature extraction.

The completely connected (dense) layers that receive feature mappings from a shared representation. These layers carry out high-level feature processing and categorisation.

The probabilities for the two classifications, "Yes" (tumour present) and "No" (tumour missing), are output by a final layer.

The following sections provide an overview of each model architecture, highlighting its features, design, and applicability to brain MRI tumour detection tasks:

AlexNet is a convolutional neural network architecture that is used because it is made of a combination of various essential parts that are required to analyse and categorise the visual data efficiently and effectively. There are five convolution layers in the network, each of which is used to extract the hierarchical features from the given input image with the help of a convolution filter, which is also called a kernel. These filers can identify a few elements, such as edges, textures, and patterns, based on the depth of the network. It also collects more intricate features from the input image. After each convolutional layer, there is a non-linear function called Rectified

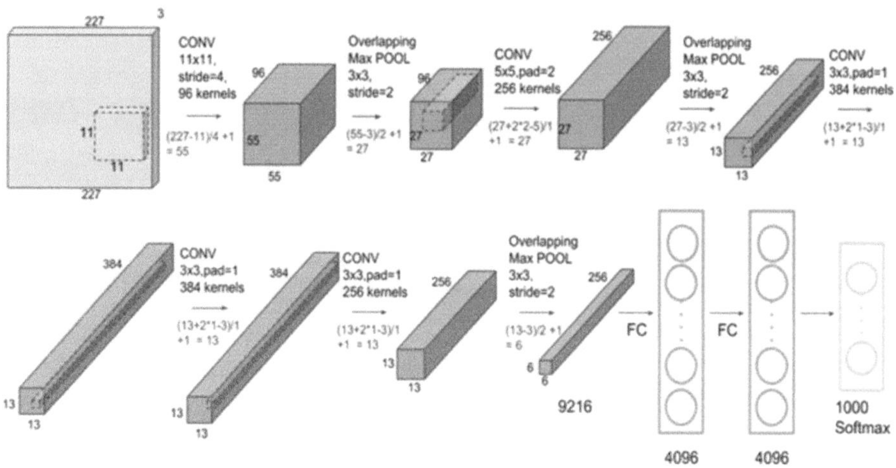

Figure 3.4 Structure of AlexNet.

Linear Unit (ReLU) to add non-linearity and also to enable the network to identify increasingly intricate patterns. Additionally, AlexNet also has three max-pooling layers that are used to select the highest value from a 2 × 2 or 3 × 3 area based on the feature map to reduce the dimensionality. This process preserves significant spatial properties while lowering the computational load.

AlexNet comprises four fully connected (FC) layers after the convolutional and pooling layers. The learnt features of the network are compressed into a 1D vector and then sent through these dense layers. The final classification decision is made by merging the features extracted by the convolutional layers with those from the fully connected layers. The output layer of AlexNet uses a Softmax layer to translate the final feature vector into class probabilities, yielding the classification result (e.g., whether a brain tumour is present). In order to normalise the feature maps and enhance generalisation, the design further incorporates Local Response Normalisation (LRN) layers following a few convolutional layers, as depicted in Figure 3.4.

The AlexNet model receives an image with dimensions of 224 × 224 × 3, but in order to properly handle the convolution processes, padding is performed during preprocessing to reduce the size to 227 × 227 × 3. Convolutional, pooling, and fully connected layers make up this planning, which makes it very effective for image classification tasks, including medical imaging applications like MRI scans that detect brain tumours.

ResNet

ResNet50 is one of the most commonly used CNNs for the identification and categorization of brain tumours from MRI images. Using a small-sized MRI dataset, the model is trained such that it will recognise the existence or absence of tumors based on the pattern and the characteristics of the image. ResNet50 is a combination

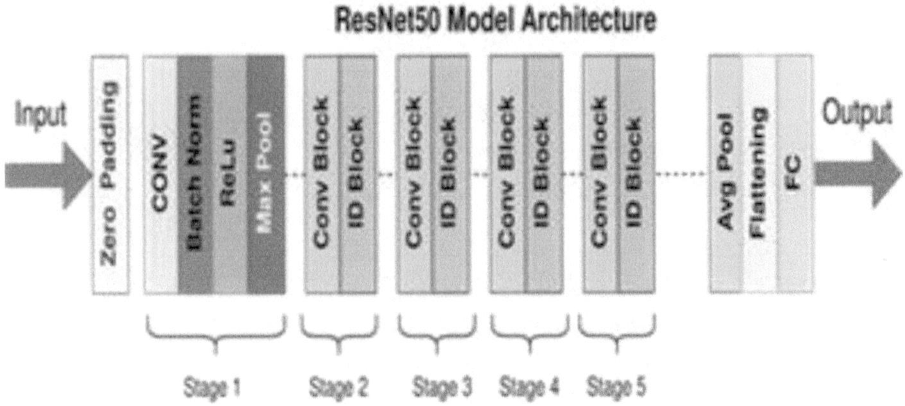

Figure 3.5 Structure of ResNet.

of convolutional layers. Pooling and fully connected layers to help the model process and extract the intricate information from the input image or data set.

ResNet50 convolutional layers are also in charge of learning the spatial hierarchies of structures, which include textures, edges, and more complicated patterns connected to tumours. The pooling layers—usually max-pooling, which assist in lowering the feature maps' dimensionality while preserving crucial data. The final classification judgement is made by combining the extracted characteristics in the fully linked layers at the culmination of the network. Residual blocks in the ResNet50 design improve gradient flow through training, allowing the network to acquire more complex representations and preventing problems like vanishing gradients illustrated in Figure 3.5.

Pre-processed brain MRI scans are used as the input for this model. Normalisation and contrast enhancement are two methods that are used to improve the dissimilarity between the tumour and the neighbouring healthy tissue. Tumour detection is facilitated by this pre-processing step, which helps the model concentrate on the pertinent features of the image. A probability distribution over the potential class labels of "tumour" (yes) or "non-tumor" (no) is the outcome of the revised ResNet50 model. This probability distribution is subjected to a threshold, which determines the final prediction of the model. The model classifies the image as "tumour" (yes) if the projected probability of a tumor presence is greater than the cutoff value; if not, it classifies the image as "non-tumor" (no). This method enables the model to produce accurate classifications for MRI image brain tumour detection.

VGG 19

VGG19 [12] Convolutional neural networks were employed in the architectural design process. Three fully linked layers, five max pool layers, sixteen convolutional layers, and the framework is composed of layers and one softmax layer.

An eminent deep CNN style for image classification tasks is VGG19, which is renowned for its ease of use and efficiency. Small 3 × 3 convolutional filters are used throughout the model to capture fine-grained information, enabling a thorough

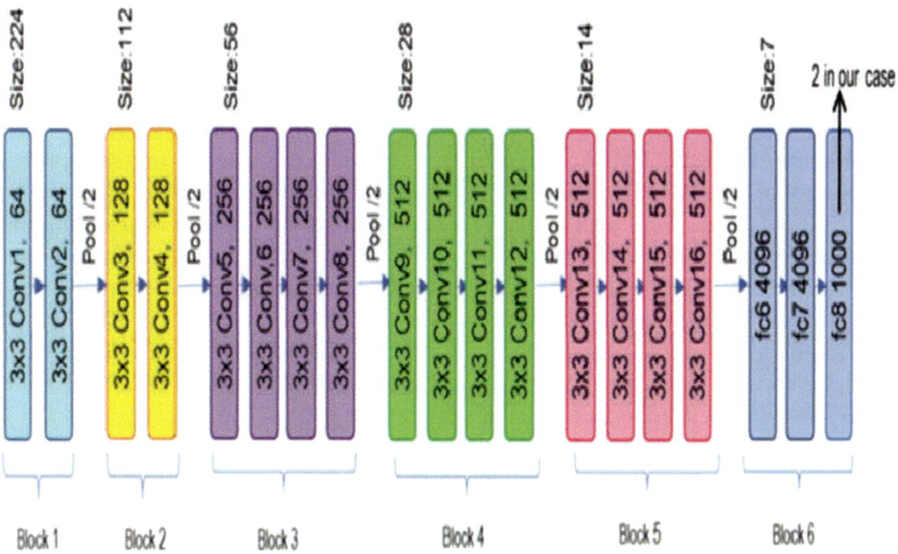

Figure 3.6 Structure of VGG 19.

analysis of the input image. As the image moves through the layers, these tiny filters allow the network to identify local patterns, including edges and textures, at various levels of abstraction. VGG19 employs max pooling layers in addition to convolutional layers to enhance translation invariance, lower computational load, and shrink the spatial size of feature maps. VGG19's pooling operation efficiently downsamples the feature maps while maintaining crucial information by using a 2 × 2 pixel window with a stride of 2. The value of the convolutional stride is 1, and maintaining spatial resolution by making sure the filters are applied to neighbouring pixels without skipping. Better feature extraction is made possible by the convolution layers' use of padding of size 1, which keeps the image dimensions from being drastically reduced, as described in Figure 3.6.

The 19 layers that make up VGG19's architecture are stacked in blocks and include 3 fully connected layers and 16 convolutional layers. The deep network permits the model to learn increasingly complicated representations of the image, with each block usually consisting of two or more convolutional layers followed by max pooling. In order to combine the learnt features for classification, the feature maps are flattened and sent through fully connected layers following the convolutional and pooling procedures.

4. Experimental Results

The creation and assessment of deep learning architecture for the categorisation and identification of brain tumours from MRI images is the main objective of this work. Brain tumours are a serious medical issue, and better treatment results are largely dependent on early discovery. However, manually identifying and categorising brain tumours from MRI scans can be laborious, arbitrary, and error-prone. To help medical practitioners make quick, reliable diagnoses, automated tumour-detection techniques are widely desired in this section.

4.1 Performance of Hyperparameters of the Transfer Learning Models

The Navoneel dataset of MRI scan pictures was used in this study to train and assess the NeuroNET model. To maintain the integrity of model validation, the training and testing datasets were kept apart. The development method included setting the required hyperparameters and using a block size of 128 for each model. The memory needs and computational effectiveness of training the deep learning model were balanced when selecting this block size.

To maximise model performance, the training process's hyperparameters were carefully chosen. Batch size, optimiser selection, and learning rate were important hyperparameters. Validation loss was used at each epoch to direct the model's learning process. A popular loss function for classification problems, especially binary classification, is cross-entropy loss, which minimizes the discrepancy between the true labels and the projected probabilities. In every epoch, the training and test sets were subjected to this loss function.

Training the model for 100 epochs allowed it to learn and make incremental adjustments to its parameters by passing the whole training dataset through the network 100 times. The model's weights were updated using the Adam optimiser. Adam is a well-known optimisation technique that is very effective for training deep neural networks, as it adjusts the learning rate for every parameter. In order to help the model converge without going overboard with the ideal solution, the learning rate was set at 0.001, a frequently used initial value.

A number of indicators were used to assess the proposed NeuroNET model performance. The model validation accuracy was 98%, meaning that 96% of the validation images were properly classified by the model. At all potential classification thresholds, the model successfully distinguished between the tumour and non-tumor classes, as evidenced in the values presented in Tables 3.1 and 3.2. A smaller loss number implies higher model performance. The validation loss, which shows the difference between the model's predictions and the actual labels on the validation set, was 0.03.

The true label predictions made by the NeuroNET model are shown in Figure 3.7, which visualizes the model's outputs. This figure provides a comparison between the actual labels and the model's predicted labels, illustrating the model's ability to precisely predict brain tumor presence in MRI images.

Table 3.1 Performance of the Proposed model with the existing model for the Br35H dataset.

Model	Accuracy of Training	Loss of Training	Accuracy of Testing	Loss of Testing
AlexNet	83%	0.38	88%	0.30
ResNet50	77%	0.42	83%	0.40
VGG19	84%	0.40	76%	0.45
Proposed Model NeuroNet	96%	0.09	98%	0.06

Table 3.2 Performance of the Proposed model with the existing model for the Navoneel dataset.

Model	**Accuracy of Training**	**Loss of Training**	**Accuracy of Testing**	**Loss of Testing**
AlexNet	69%	0.60	72%	0.57
ResNet50	72%	0.53	70%	0.52
VGG19	71%	0.9	74%	0.8
Proposed Model NeuroNet	88%	0.25	91%	0.24

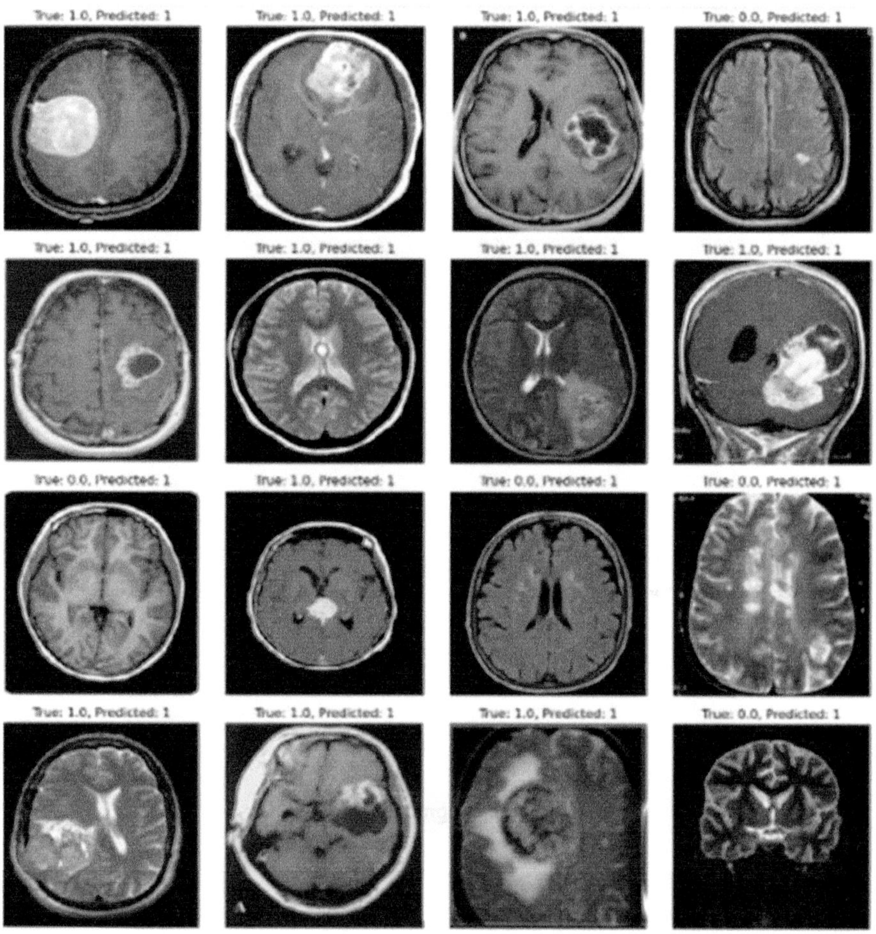

Figure 3.7 True Label and predicted label by NeuroNET model.

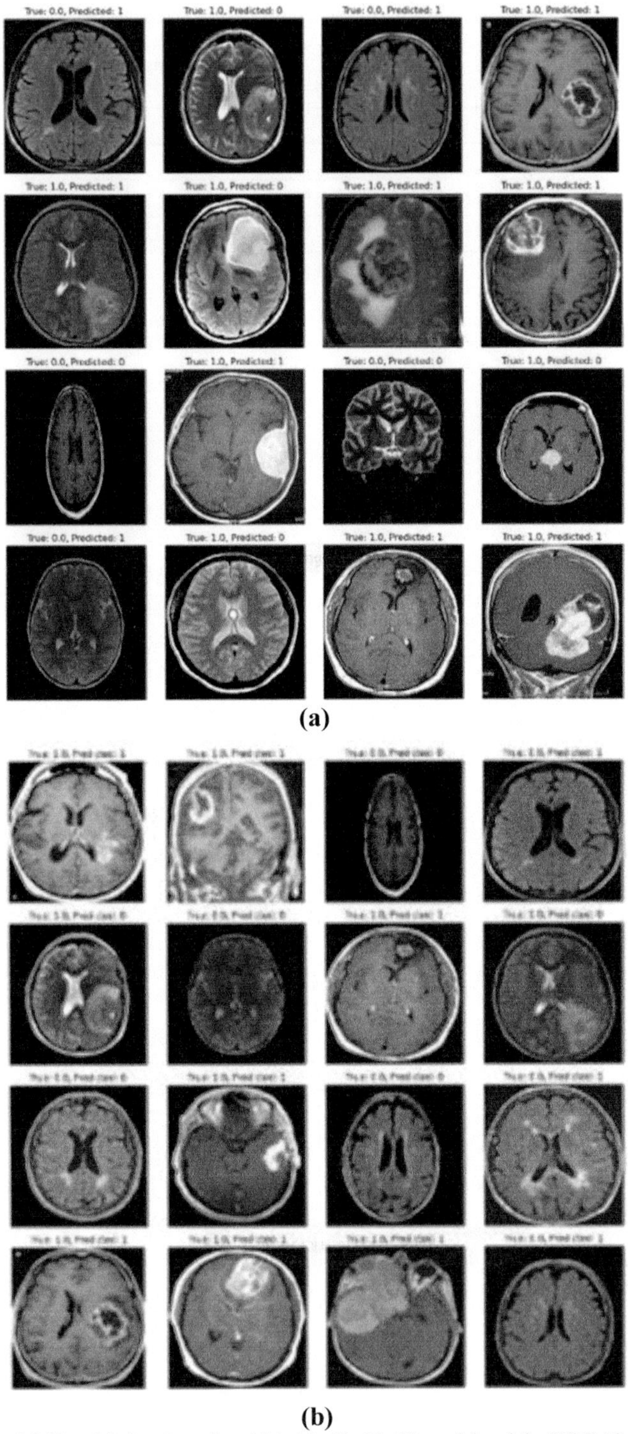

Figure 3.8 True label and predicted label by ResNet50 model and the VGG 19 model.

Figure 3.8(a) shows how the input MRI images are categorised using the extracted features and offers a thorough overview of the process flow for training and implementing the CNN-based brain tumour detection ResNet 50 model.

Figure 3.8(b) shows how the input MRI images are categorised using the extracted features and offers a thorough overview of the process flow for training and implementing the CNN-based brain tumour detection VGG 19 model.

4.2 Evaluating Existing Methodologies in Context

Using both training and testing datasets, the proposed framework has been thoroughly validated by an independent assessment against a number of reputable pre-trained models, including VGG19, AlexNet, and ResNet50. These models were chosen because they are widely used and perform well in a variety of picture classification tasks, such as the detection of brain tumours. For ease of comparison, the performance characteristics of these current models and the anticipated framework are shown in Table 3.1.

The projected approach continuously achieves improved results compared to the current models in terms of testing accuracy. The proposed framework showed improved accuracy, indicating that it is better at generalising and making predictions on unseen MRI brain scan data, even though more conventional models like VGG19, AlexNet, and ResNet50 have shown in Figure 3.9 encouraging results in a variety of applications.

The CNN model, which serves as the foundation for the suggested framework, integrates cutting-edge methods such as improved feature extraction, optimised architectures, and enhanced training procedures. This enables it to process MRI pictures efficiently and identify intricate patterns that may be signs of brain tumours. The suggested model showed a significant improvement in performance across

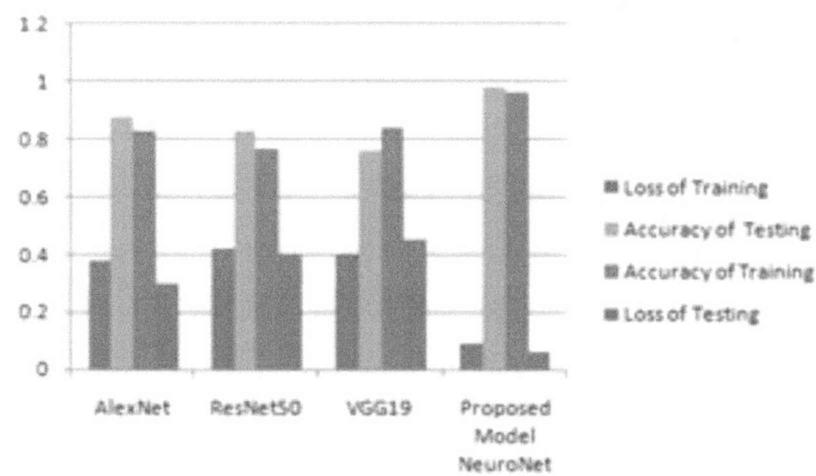

Figure 3.9 Comparison between AlexNet, ResNet 50 model, and VGG 19 model with the Proposed NeuroNET model.

several evaluation criteria when compared to existing models, which depend on more typical convolutional layers and activation functions. All things considered, the findings show that the proposed framework has a distinct advantage over current models, making it a viable option for precise and effective brain tumour identification from MRI pictures.

4.3 Comparison Evaluations for the Model

The investigation of the exactness graph revealed that the CNN-based architecture achieved better results than the other models when the CNN models' performances were compared shown in Figure 3.10. The CNN model performed better in both the training and validation stages, as this graph that contrasts the validation accuracy with the training accuracy makes evident. For real-world applications like brain tumour diagnosis in MRI images, the model's consistent and high validation accuracy showed that it generalised well to unseen data.

Furthermore, during the training process, the training and validation precision curves stayed near one another, as shown in Figure 3.11. This suggests that neither overfitting nor underfitting was a major problem. The model performs poorly on new data when it overfits, which occurs when it learns too much from the training set and captures noise or unimportant patterns. Conversely, underfitting occurs when the model performs worse than expected because it cannot capture the main patterns in the data. In this instance, the model appears to have learnt the required features without overfitting or underfitting, as indicated by the near alignment between training and validation accuracy.

Overall, the research demonstrates that the proposed NeuroNET model maintained good accuracy on both the training and validation sets without exhibiting any discernible overfitting or underfitting, demonstrating a balanced learning

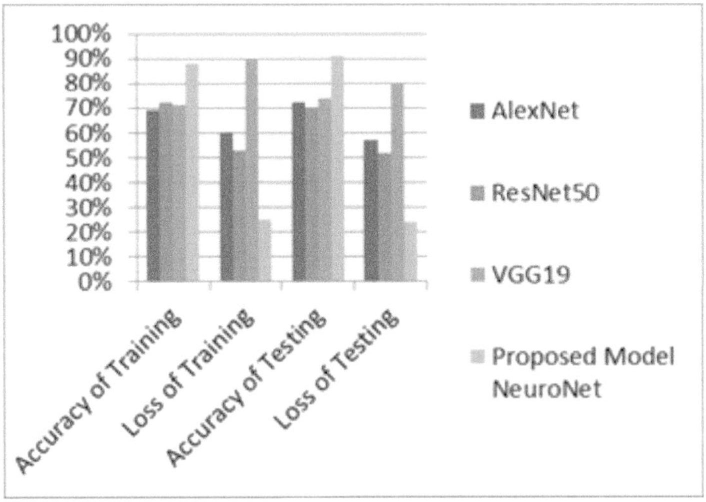

Figure 3.10 Comparative analysis of the model's performance.

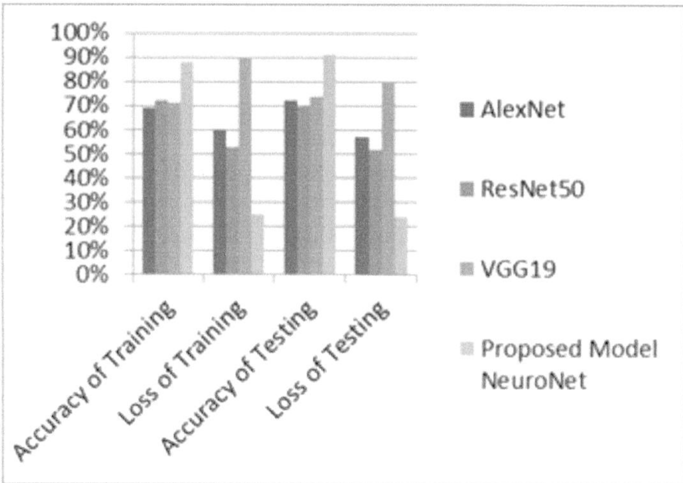

(a)

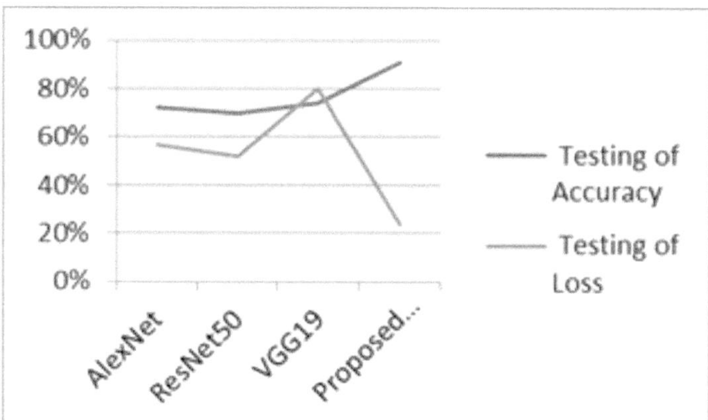

(b)

Figure 3.11(a) and (b) Performance analysis of AlexNet, ResNet 50 model, and VGG 19 model with the Proposed NeuroNET model.

process. This demonstrates even more how dependable and successful the suggested model for classifying brain tumours is.

5. Conclusion

This study investigated the application of transfer learning algorithms for brain tumour identification and classification using MRI images. The performance of four different architectures—AlexNet, VGG19, ResNet50, and the suggested NeuroNET model—was assessed using datasets from brain tumours. By attaining an 82% testing accuracy and a lower testing loss of 0.47, NeuroNET outperformed the other models. This suggests that NeuroNET is a more reliable method for classifying brain tumours from MRI images than the current models since it performs better in terms of both precision and generalisation.

The NeuroNET potential as a useful model for practical medical imaging applications is highlighted by its capacity to attain more precision with less loss. The architecture's optimised layers and the use of cutting-edge transfer learning techniques may be responsible for this enhanced performance, as they enable it to capture the intricate patterns connected to brain tumours.

In the future, this approach will be evaluated further by adding more MRI scan pictures to the dataset in order to increase the generalisability and robustness of the model. To improve the model's performance, other issues with deep learning architectures, like model overfitting, training effectiveness, and computing cost, will also be addressed. Future investigations will focus on refining and fine-tuning the NeuroNET model while investigating fresh approaches and tactics for more precise and effective brain tumour identification and categorisation.

References

[1] Chattopadhyay, Arkapravo, Maitra and Mausumi. (2022). MRI-based brain tumor image detection using CNN based deep learning method. Neuroscience Informatics 2: 100060. 10.1016/j.neuri.2022.100060.

[2] Mahmud, M.I., Mamun, M. and Abdelgawad, A. (2023). A deep analysis of brain tumor detection from MR images using deep learning networks. Algorithms 16(4): 176. https://doi.org/10.3390/a16040176.

[3] Md Ishtyaq Mahmud, Muntasir Mamun and Ahmed Abdelgawad. (2023). A deep analysis of brain tumor detection from MR images using deep learning networks. Algorithms 16(4): 176; https://doi.org/10.3390/a16040176, Published: 23 March 2023, Special Issue Machine Learning in Healthcare and Biomedical Application II.

[4] Chetana Srinivas, Nandini Prasad K. S., Mohammed Zakariah, Yousef Ajmi Alothaibi, Kamran Shaukat, B. Partibane et al. (2022). Deep transfer learning approaches in performance analysis of brain tumor classification using MRI images. Journal of Healthcare Engineering vol. 2022, Article ID 3264367, 17 pages. https://doi.org/10.1155/2022/3264367.

[5] Qureshi, Shahzad, Raza, Shan e Ahmed, Hussain, Lal, Nour, Mohamed, Rehman, Aziz, Al-Wesabi, Fahd et al. (2022). Intelligent ultra-light deep learning model for multi-class brain tumor detection. Applied Sciences 12: 3715. 10.3390/app12083715.

[6] Zahoor, Qureshi, Shahzad, Bibi, Sameena, Khan, Saddam and Khan, Asifullah. (2022). A new deep hybrid boosted and ensemble learning-based brain tumor analysis using MRI. Sensors 22: 7. 10.3390/s22072726.

[7] Arabahmadi, Mahsa, Farahbakhsh, Reza and Rezazadeh, Javad. (2022). Deep Learning for Smart Healthcare: A Survey on Brain Tumor Detection from Medical Imaging. Sensors (Basel, Switzerland). 22. 10.3390/s22051960.

[8] Tandel, S.G., Biswas, M., Kakde, G.O., Tiwari, A., Suri, S.H., Turk, M. et al. (2019). A review on a deep learning perspective in brain cancer classification. Cancers (Basel) 2019 Jan 18; 11(1): 111. doi: 10.3390/cancers11010111. PMID: 30669406; PMCID: PMC6356431.

[9] Gokila Brindha, P., Kavinraj, M., Manivasakam, P. and Prasanth, P. (2021). Brain tumor detection from MRI images using deep learning techniques. IOP Conference Series: Materials Science and Engineering 1055: 012115. 10.1088/1757-899X/1055/1/012115.

[10] Chithra, P.L. and Dheepa, G. (2020). Di-phase midway convolution and deconvolution network for brain tumor segmentation in MRI images. International Journal of Imaging Systems and Technology (1–13): 22407.

[11] Chithra, P.L. and Janes Pushparani, S. (2019). SURF points versus SIFT points in identification of medicinal plants. International Journal of Engineering and Advanced Technology 9(Issue 2) IF-1.0: 602–607, December 2019, IF-1.0, ISSN 2249-8958.

[12] Khattar, Anuradha and Quadri, Syed. (2022). Generalization of convolutional network to domain adaptation network for classification of disaster images on twitter. Volume 81, Sep 2022.
[13] Aggarwal, M., Tiwari, A.K., Sarathi, M. and Anchit Bijalwan. (2023). An early detection and segmentation of brain tumors using a deep neural network. BMC Med Inform Decis Mak 23: 78. https://doi.org/10.1186/s12911-023-02174-8.
[14] Dheepa, G. and Chithra, P.L. (2022). Brain tumor detection in MRI images using input cascaded CNN with wavelet features. NeuroQuantology 20(19): 2660.
[15] Salçin, K. (2019). Detection and classification of brain tumours from MRI images using faster R-CNN. Tehnički glasnik 13(4): 337–342.
[16] Sajid, S., Hussain, S. and Sarwar, A. (2019). Brain tumor detection and segmentation in MR images using deep learning. Arab J. Sci. Eng. 44: 9249–9261. https://doi.org/10.1007/s13369-019-03967-8.
[17] Ayadi, W., Elhamzi, W., Charfi, I. and Atri, M. (2021). Deep CNN for brain tumor classification. Neural Process. Lett. 53: 671–700.
[18] ZainEldin, H., Gamel, S.A., El-Kenawy, E.-S.M., Alharbi, A.H., Khafaga, D.S., Ibrahim, A. et al. (2023). Brain tumor detection and classification using deep learning and sine-cosine fitness GreyWolf optimization. Bioengineering 10: 18.
[19] https://www.kaggle.com/navoneel/datasets.
[20] https://www.kaggle.com/datasets/ahmedhamada0/brain-tumor-detection.
[21] Srikanth, B. and Suryanarayana, S.V. (2021). Multi-class classification of brain tumor images using data augmentation with deep neuralnetwork. Mater. Today Proc. [Google Scholar]
[22] Musallam, A.S., Sherif, A.S. and Hussein, M.K. (2022). A new convolutional neural network architecture for automatic detection of brain tumors in magnetic resonance imaging images. IEEE Access 10: 2775–2782. [Google Scholar].
[23] Andrew, A. and Santoso, H. (2022). Compare VGG19, ResNet-50, Inception-V3 for review food rating. Sink. J. Dan Penelit. Tek. Inform. 7: 845–494 [Google Scholar].
[24] Liu, M., Yi, M., Wu, M., Wang, J. and He, Y. (2022). Breast pathological image classification based on VGG16 feature concatenation. J. Shanghai Jiaotong Univ. (Sci.) 27: 473–484 [Google Scholar] [CrossRef].
[25] Bansal, M., Kumar, M., Sachdeva, M. and Ajay Mittal. (2023). Transfer learning for image classification using VGG19: Caltech-101 image data set. J. Ambient Intell. Human Comput. 14: 3609–3620. https://doi.org/10.1007/s12652-021-03488-z.
[26] Wanni Xu, You-Lei Fu and Dongmei Zhu. (2023). ResNet and its application to medical image processing: Research progress and challenges, Computer Methods and Programs in Biomedicine 240: 107660, ISSN 0169-2607, https://doi.org/10.1016/j.cmpb.2023.107660.
[27] Nayak, D.R., Padhy, N., Mallick, P.K., Zymbler, M. and Kumar, S. (2022). Brain tumor classification using dense efficient-net. Axioms 11: 34 [Google Scholar].
[28] Gaikwad, N. and Vijayalakshmi, S. (2024). An examination of methodological approaches for segmentating fetal brain MRI images – Analysis. 2024 International Conference on Trends in Quantum Computing and Emerging Business Technologies, Pune, India, 2024, pp. 1–6, doi: 10.1109/TQCEBT59414.2024.10545035.
[29] Gayathri, S.P., Siva Shankar and Vijayalakshmi Subrmanian. (2023). Machine learning approaches towards medical images. Structural and Functional Aspects of Biocomputing Systems for Data Processing, edited by U. Vignesh, et al., IGI Global Scientific Publishing, pp. 124–145. https://doi.org/10.4018/978-1-6684-6523-3.ch006.
[30] Durgadevi, P. and Vijayalakshmi, S. (2023). Foetal brain extraction using mathematically modelled local foetal minima, International Journal of Biomedical Engineering and Technology (IJBET) Vol. 42, No. 3.

Chapter 4

Highlighting Lung Cancer Using 3D Grad-CAM in Pivot Point CNN

P. Bhavani and *P.L. Chithra**

1. Introduction

According to the American Cancer Society (2023), cancer has grown to be an essential issue for the public, and millions of people worldwide die silently from cancer. As per the World Health Organization's (WHO) survey report, cancer was predicted to cause 10 million fatalities in the second-highest mortality [1]. Anybody can get cancer at any age, and it can affect any part of the body. There are over 100 distinct forms of cancer, each with unique traits and behaviors, as outlined by U.S.D.o.H.a.H. Services et al. [2].

The most common types of the disease, according to WHO (2023), are liver, stomach, colorectal, lung, and breast cancers. The origin of LC is the lungs [3]. Breathing is the process by which the body exchanges oxygen and carbon dioxide, and the two organs that comprise the lungs are located in the chest. It was estimated that 1.80 million lung cancer-related deaths occurred in 2021 [4]. Lung cancer can be classified as either small-cell or non-small-cell [5].

This lung cancer classification technique determines the stages of lung cancer based on tumor location, size, and shape, using lung imaging. The analysis of medical image data, such as X-rays and internal body images, can be used to diagnose patients using MRI (Magnetic Resonance Imaging) and CT (Computerised

Department of Computer Science, University of Madras, Chennai – 600 025, India.
Email: bhavanipandiyan.91@gmail.com
* Corresponding author: chitrasp2001@yahoo.com

DOI: 10.1201/9781003635178-5

Tomography) imaging [6]. The radiologist examines the CT and MRI images of the lungs to identify and detect the abnormal tumor [7]. Two methods for analyzing medical images, specifically lung cancer images, that use machine learning algorithms to identify cancer based on characteristics. Creating deep learning models is an additional method for image analysis. The most popular convolutional Neural Networks (CNNs) are models for image and video classification problems, while there are many more variations of DL models [8].

CNNs are especially advantageous because they can extract complex information from raw data through hierarchical learning representation. Without being hampered by manual feature encoding, a CNN automatically identifies and extracts notified features from the given image. When images share the same attributes, a CNN can still categorize them effectively even when they differ greatly from the training examples. CNNs require a substantial amount of data to function well. Obtaining a large dataset may be challenging [9].

EfficientNets' transfer learning offers several benefits that can help with frequent issues in analyzing medical images. The significance of EfficientNets lies in their ability to extract accurate features from lung images, a well-established property. The EfficientNet models are pre-trained to extract relevant features from large-scale datasets (such as ImageNet) and then apply transfer learning to apply this understanding to the lung cancer segmentation challenge. This may enhance the segmentation process's resilience and accuracy [10–13].

This research has four categories: Section 2 indicates related works on highlighting tumor regions to predict lung cancer; Section 3 points to the proposed 3D Pivot Point of network architecture; Section 4 analyzes the experimental solutions and compares them with implementing existing techniques to IQ-OTH/NCCD datasets; and finally, Section 5 generates conclusions.

2. Related Works

Early identification and treatment of lung cancer are critical to averting fatalities. Because malignant cells are growing abnormally in one or both lungs, LC is thought to be a fatal disease that, if left untreated, could spread to other human organs. Consequently, a high-performing computer-aided diagnostic (CAD) framework is needed to accurately and automatically identify and classify more lung cancer cases. Various medical imaging techniques, including CT scans, X-rays, histopathology images, and sputum microscopy smear images, can be used to identify, segment, and classify malignant lung nodules [14].

Kareem et al. proposed that, before deep learning became more advanced, a variety of image processing techniques were used for manual feature extraction, including Adaptive bilateral filtering, fuzzy C-means clustering, particle swarm optimization, top-hat transformation, median noise removal, segmentation using the marker watershed algorithm, and morphological operations. A proposed CAD-based method for classifying malignant and non-malignant lung cancer also used the retrieved feature vectors, which were subsequently fed to various machine-learning algorithms for segmentation and classification [15].

Raza et al. implemented stochastic deep learning architectures that yield distinct outcomes for every prediction, and the deep learning model's output is characterized by randomness and uncertainty. Furthermore, because the dataset is highly volatile, there is a risk of overfitting on the training set. Several research studies in various fields, including medical imaging, have employed ensemble learning to overcome the aforementioned problems [16].

Humayun et al. created three groups—normal, malignant, and benign—using a multi-class approach based on transfer learning to categorize lung cancer from CT images. By using an image augmentation technique, they instinctively increased the samples from the datasets. With the publicly accessible IQ-OTH/NCCD lung cancer dataset, we rebuilt three existing architectures of Xception, VGG16, and VGG19 [16].

Al-Yasriy et al. present architectures for the Xception, VGG-16, and VGG-19 networks that achieved total accuracies of their projected transfer learning: 98.05%, 97.4%, and 98.83%, respectively. Comparably, the researchers carried out AlexNet's architecture transfer learning process on CT scan images trained in the IQ-OTH/NCCD lung cancer dataset. Learning of AlexNet [17–19].

In addition to classification, some research has focused on identifying and separating healthy tumors from lung images obtained from CT or X-ray scans using different transfer-learning-based algorithms [20].

These are the principal contributions of this proposed work:

- This research work developed the proposed Pivot Point CNN Architecture for lung cancer classification.
- The proposed 3D Pivot point network addresses the extreme disparity data challenges by overcoming the dispersion of the data through image augmentation.
- To perform many experiments, many experiments were carried out using the existing dataset, executing image augmentation to assess the impact of this process on the proposed model's performance.
- To show the effectiveness of the proposed Pivot Point CNN architecture over other classification models, the proposed work compared its implementation duration and computational challenges with those of other cutting-edge techniques.
- The proposed model, Pivot Point CNN architecture, shows better results than existing techniques and uses CT scan images to identify lung cancer. This invention only concentrated on CNN architectures and techniques of Grad-CAM, which relate specifically to the proposed, even though there is much related work on automated medical image segmentation.

3. Proposed Methodology

In this proposed 3D Pivot Point network, the feature map is created separately, by convolving each input image with two different kernels. The convolutional and deconvolutional layers process the global average of convolved features. After processing, the value is fed into the softmax layer to classify areas with and without

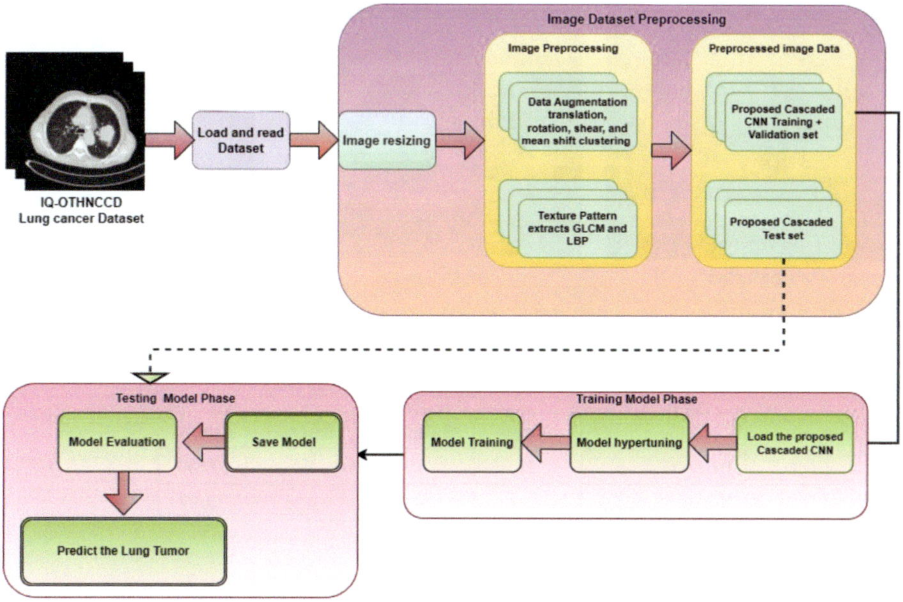

Figure 4.1 3D Lung Cancer Proposed Methodology Workflow.

tumors into classes. Finally, the softmax layer identifies tumor regions and highlights the tumor positions. Figure 4.1 shows the proposed methodology for 3D Lung Cancer.

3.1 Lung Cancer Dataset

According to Alyasriy and Muayed (2020), the IQ-OTH/NCCD lung cancer dataset taken from the "National Centre for Cancer Diseases/Iraq-Oncology Teaching Hospital is used for the tests. More than three months were spent on the dataset, which was collected in 2019 from the National Centre for Cancer Diseases and the Iraq-Oncology Teaching Hospital. Patients' CT scans, whether healthy or at various stages of lung cancer, are the dataset's components [21]. Two radiologists and one oncologist annotated the dataset. With respect to age, gender, education level, place of residence, and way of life, the dataset includes CT lung cancer images taken of humans in 1,097 images instead of 110 slices, as shown in Figure 4.2, which displays the categories of the lung cancer patients' cases, like benign lung cancer, malignant lung cancer, and normal lung.

Examining 110 slices, 40 were cancerous, 15 were benign, and 55 were normal. The 512 × 512 DICOM CT scans utilized in the research were acquired initially; however, they were converted to JPEG format for convenience. The Kaggle repository [19] has the dataset IQ-OTH/NCCD accessible to the general public. Table 4.1 displays the dataset's breakdown by class-wise category.

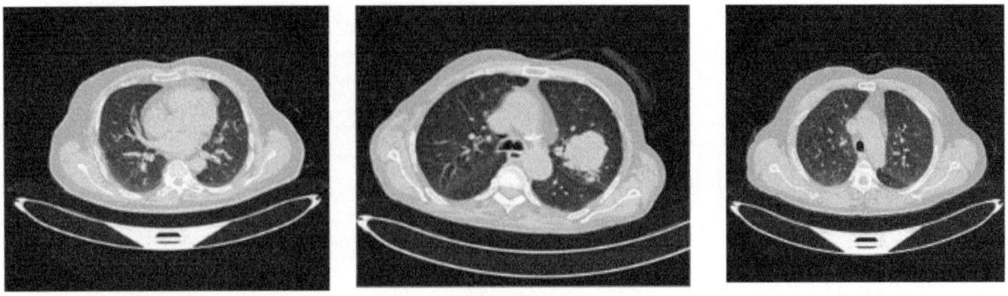

(a) Benign (b) Malignant (C) Normal

Figure 4.2 Lung cancer dataset (IQ – OTH/NCCD) in each class for Benign, Malignant, and Normal.

Table 4.1 Collection of IQ – OTH/NCCD Dataset.

Class	Benign	Malignant	Normal	Total
Patients	15	40	55	110
No. of Samples	120	561	416	1,097

3.2 Lung Cancer Data Pre-Processing

The proposed 3D pivot point model was trained and tested before performing the preprocessing techniques. Each class's images are rearranged for unbiased training and divided into an 80:20 proportion of the training and testing sets. Accordingly, 20% of lung images are analyzed for the testing set, and 80% of lung images for the training data set, which is used for training and evaluating the model on a possible test case. Table 4.2 presents the dataset's class-wise distribution following the train-test split.

Table 4.2 Lungcancer Dataset.

Lung Data Class	Splitting Dataset	No. of Collection	Total
Benign	Train set	96	876
Malignant		448	
Normal		332	
Benign	Test set	24	221
Malignant		113	
Normal		84	

3.3 Lung Cancer Image Augmentation

The datasets for each class are improved by applying image augmentation to the original image samples. To train the 3D Pivot Point CNN architecture, the sample of the lung CT 1190 dataset is inadequate. Data augmentation also improves the model's adaptability and prevents overfitting. By applying different image alteration techniques, including cropping, translation, rotation, shearing, and horizontal and vertical flips, this widely used method increases the amount of automatically

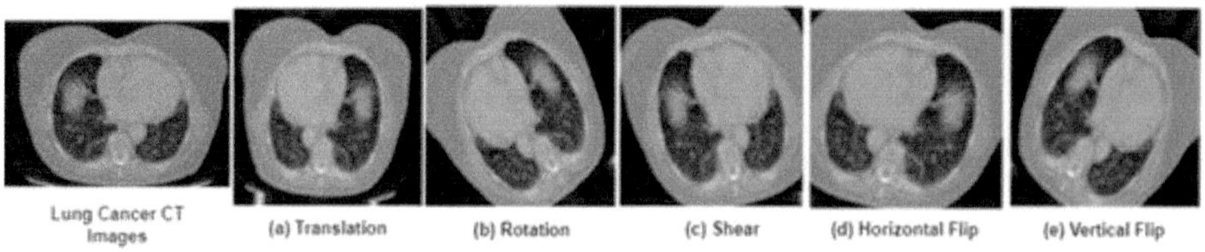

Figure 4.3 Image Augmentation of CT Lung Cancer Images.

generated datasets. The benign, malignant, and normal lung image classes are enhanced using these three image data augmentation approaches, which partially help each class of the sample balance dataset. Augmentation of the image is limited to CT image slices inside a train set, as shown in Figure 4.3.

3.4 3D Lung Cancer Pivot Point CNN

The proposed 3D Lung Cancer Pivot Point network is implemented in Figure 4.4. This network has been designed to have eight MaxPooling3D layers and eight convolutional layers. The size and shape of the filters with every new layer, the inputs vary. The size of the image (5 × 5 × 5) (width, height, depth, number of channels) makes up the lung input shape of the 3D image (64, 512, and 512) to the CNN; 32 filters with a kernel measuring 7 × 7 × 7 make up the first layer of 3D input. The MaxPooling3D layer featured a 2 × 2 × 2 proportion with a stride of 2, and the input shape determined the number of filters and the quantity of the input filter between it and the first 3D convolution layer. The data volume is decreased via MaxPooling. Sixty-four identical filters, each measuring three by three by three, are included in the 3D second convolution layer exactly like in the first. The second layer of 3D convolutional is separated by a MaxPooling of 3D of the same dimensions, 2 × 2 × 2. Filter size of 128 (3 × 3 × 3) is obtained by combining the third and fourth convolution layers of 3D after the global average pooling of two feature values and the Maxpooling3D layer that came before it. A MaxPooling with the same dimensions as the previous MaxPooling layers—2 × 2 × 2—follows the fourth convolution 3D layer. The final phase involves using the two 3D convolution layers, one containing 512 filters with a size of 1 × 1 × 1 and the other with 512 filters of the same size. After the 256 layers with 512 filters, there is a final 2 × 2 × 2 MaxPooling of the 3D layer. In the background, each 3D MaxPooling layer is undergoing batch normalization. The network termination has a thick, flattened layer. The number 11 is taken as-is by the single dense layer. The filter size was chosen with the output of the filters in mind, wherein adherence to the integer output was necessary. The optimization technique "Adam," Its default learning rate is 0.1, and it was this architecture that was chosen. Weight was decreased, and the Pivot Point GradCAM model was used in the final convolution layer as a result of varying batch sizes and epoch settings.

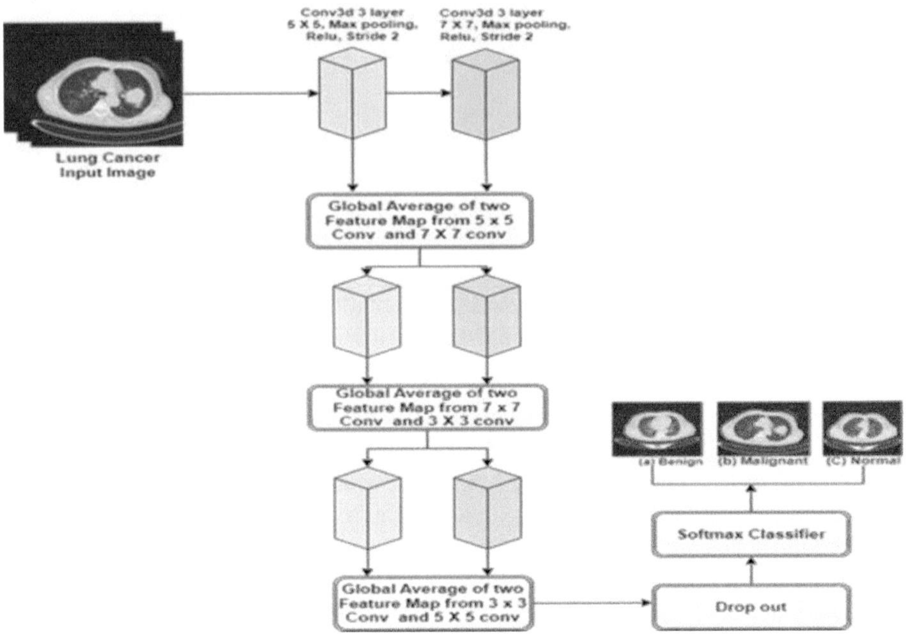

Figure 4.4 Proposed 3D Lung Cancer Pivot Point CNN.

4. Experimental Results and Discussion

4.1 Performance of Evaluation Measures and Implementation

A deep learning model's effectiveness is determined by evaluating it using evaluation measures, which are quantifiable measurements. Evaluation metrics are used to compare how well one model or algorithm performs with another on a given task, find areas for improvement, and determine how effective a model or algorithm is at solving a certain problem. The evaluation metrics are recall, precision, accuracy, and F1-Score, as given by the equations below.

$$Accuracy = \frac{(TP + TN)}{(TP + TN + FP + FN)} \tag{1}$$

$$F1\ score = \frac{2 * (Precision \times Recall)}{(Precision + Recall)} \tag{2}$$

$$Precision = \frac{TP}{(TP + FP)} \tag{3}$$

$$Recall = \frac{TP}{(TP + FN)} \tag{4}$$

4.2 Comparative Analysis Existing Methodology

Compared with the existing methodology of lung image pre-processing techniques and fine-tuning, the proposed Pivot Point CNN model allowed the model to generate exceptional outcomes using image classification methods. In order for features to be extracted from the lung images, this model combines convolutional and pooling layers, building on the characteristics obtained by the layer that came before it. The model has to be able to learn intricate models of lung images to categorize lung images appropriately. Interestingly, this model yields 100% precision and 99.20% overall accuracy, indicating that it can distinguish between benign and malignant lung images with reasonable accuracy, as shown in Table 4.3, and model accuracy displayed in Figure 4.5. Furthermore, the model's 100% specificity shows that it can recognize normal lung images. The high efficiency of the proposed Pivot Point architecture feature layers has been demonstrated in Figure 4.6.

Table 4.3 Comparative Analysis of the Pivot Point CNN with Existing Methodology.

Pivot point CNN Model	70:30	80:20	90:10
Model Accuracy	98	99.2	89.1
Precision	97.1	96	85
Recall	94.13	92	74
F1 Score	95.1	94	87.9

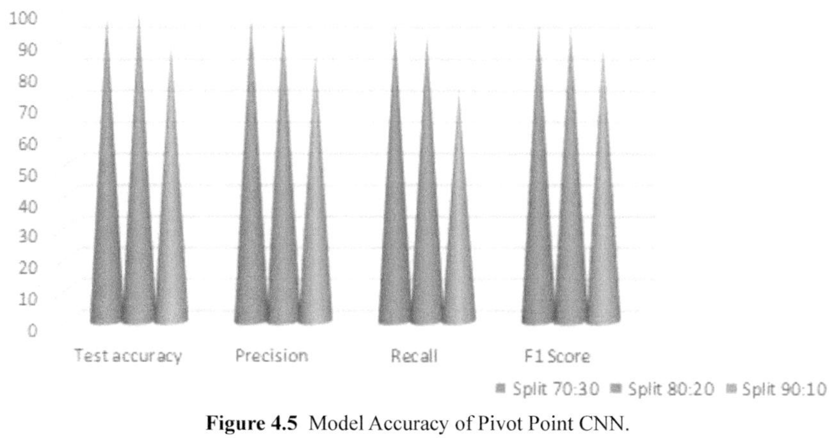

Figure 4.5 Model Accuracy of Pivot Point CNN.

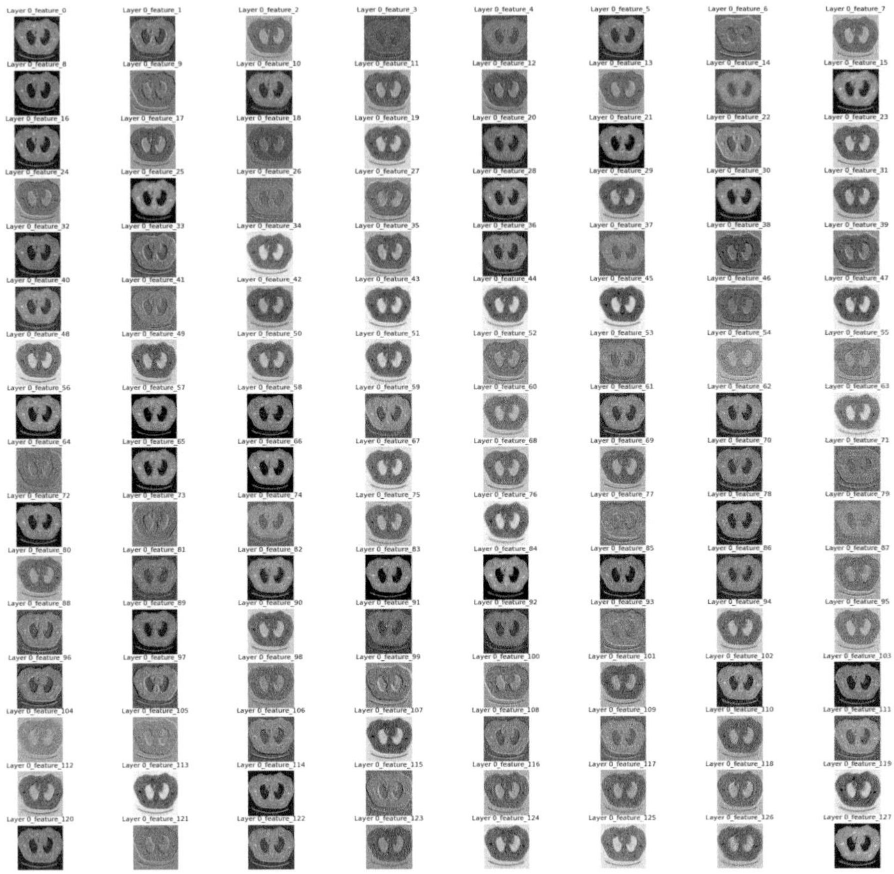

Figure 4.6 Feature Layer Visualization of Pivot Point CNN.

4.3 Comparative Analysis of Benchmark Dataset

The proposed Pivot Point CNN model has been shown to perform outstanding outcomes on lung CT image classification benchmark datasets. This Pivot Point CNN model is an effective model for predicting lung cancer on images from CT scans compared with previously trained EfficientNet and ResNet. The proposed model produced outstanding results that can be classified into lung image tests. Benign, malignant, and normal categories are applied to the dataset. Deep learning and data augmentation are employed in the proposed model to extract the microscopic tumor from CT images, as implemented in Table 4.4.

Table 4.4 Comparative Analysis of the Pivot Point CNN with Existing Methodology.

Reference	Methodology	Dataset	Accuracy
Al-Yasriy et al. [17]	AlexNet using CNN	CT imaging data	93.54%
Chen et al. [23]	CNN with embedded NLP	CT imaging data	88.00%
AL-Huseiny et al. [14]	GoogLeNet DNN	IQ-OTH/NCCD	94.38%
Narin and Onur 2022 [24]	AlexNet with ResNet	IQ-OTH/NCCD	98.58%
Lei Lyu et al. [10]	Ensemble of DenseNet. DCNN, VGG, and AlexNet	CT imaging data	99.00%
Abida et al. [25]	Transfer learning based VGG-16 and Inception V3	CT imaging data	99.8% accuracy on static data 75% accuracy on dynamic data
Mamoona et al. [26]	Ensemble approach TSCNN	CT scan image from the Luna dataset	98.83% on VGG16 98.05% on VGG19, and 97.4% on Xception
Rehan Raza et al. [22]	Transfer learning of Fine-tuned EfficientNetB1	IQ-OTH/NCCD	99.10%
	Proposed Pivot Point CNN	IQ-OTH/NCCD	99.20%

4.4 Performance of Pivot Point GradCAM Visualization

The effectiveness of the 3D Pivot Point Grad-CAM CNN model in comparison to the ResNet and EfficientNet models. As evidenced by Figures 4.6(a) and 4.6(b), representing the outcomes of the EfficientNet and ResNet models, respectively, and Figure 4.7, representing the Grad-CAM visualization, the suggested model has outperformed the other models based on this comparative analysis.

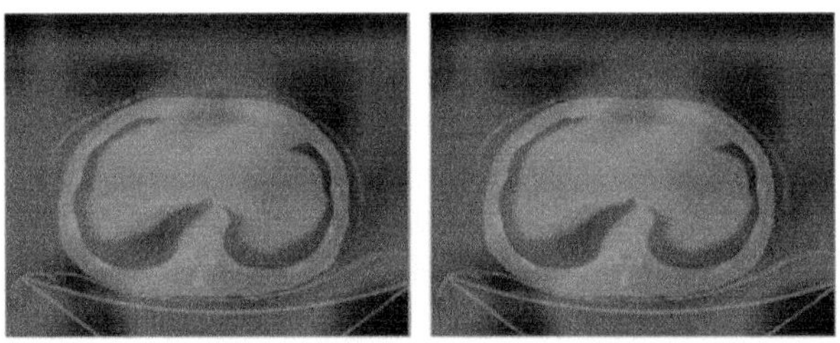

Figure 4.7(a) Outcomes from EfficientNet.

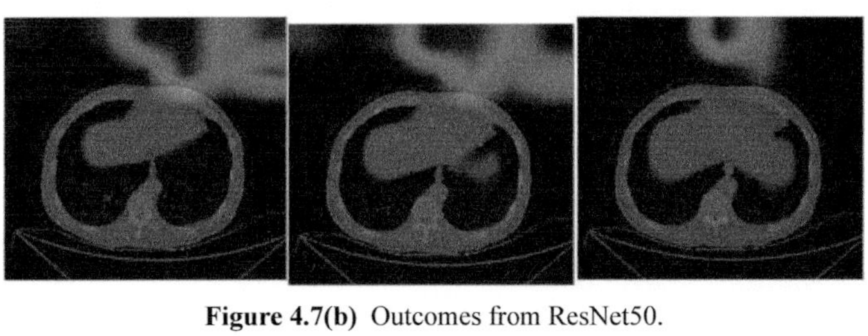

Figure 4.7(b) Outcomes from ResNet50.

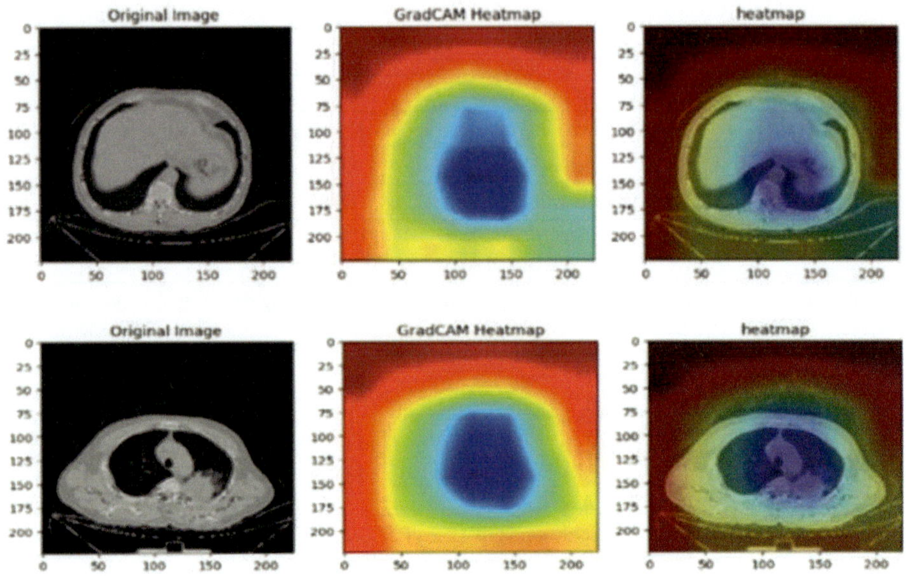

Figure 4.7(c) GradCAM Visualization of Pivot Point CNN.

5. Conclusion

The proposed research focused on the effectiveness and potential of the Pivot Point CNN in extracting key characteristics from lung cancer CT images. This technique essentially decreases training time and processing resources while improving accuracy. The proposed Pivot Point CNN achieves an amazing accuracy of 99.20%, outperforming CNN designs in terms of efficiency and accuracy. These findings highlight the tumor location for automated lung cancer categorization and diagnosis. However, using a bigger dataset might improve the suggested model's performance even more. This proposed model, compared with EfficientNet and ResNet, could be evaluated to determine its usefulness in categorizing lung cancer. This demonstrates that an automatic learning algorithm can recognize lung features in a CT scan. Future research will attempt to fully understand the development of 3D models by utilizing massive datasets produced from different medical imaging sources.

Acknowledgment

MHRD RUSA 2.0 is funding this research project. Biomaterials: The importance of Nano-HAp in cancer diagnosis and treatment: Schema and System Design under Functional Synthetic Material for Biomedical Applications Theme-2, which the University of Madras has funded.

References

[1] American Cancer Society, I. (2023). What is lung cancer? Available: http://www.cancer. org/cancer/what-is-cancer.html.

[2] U.S.D.o.H.a.H. Services, Health, N.I.o. and Institute, N.C. (2023). Cancer treatment. Available: http://www.cancer.gov/about-cancer/treatment.

[3] WHO. (2023). Cancer. Available: http://www.who.int/health-topics/cancer#tab=tab_1.

[4] Hiremath, I.S., Goel, A., Warrier, S., Kumar, A.P., Sethi, G. and o.C.P. Garg, M.J.J. (2022). The multidimensional role of the WNT/β-catenin signaling pathway in human malignancies. 237: 199–238.

[5] Ge, F., Zhu, Y.-H., Xu, J., Muhammad, A., Song, J., Yu, D.-J.J.C. et al. (2021). MutTMPredictor: Robust and accurate cascade XGBoost classifier for prediction of mutations in transmembrane proteins. 19: 6400–6416.

[6] Wang, S., Dong, L., Wang, X. and Wang, X.J.O.M. (2020). Classification of pathological types of lung cancer from CT images by deep residual neural networks with transfer learning strategy. 15: 190–197.

[7] Ge, F., Zhang, Y., Xu, J., Muhammad, A., Song, J. and Yu, D.-J. (2022). Prediction of disease associated ssnps by integrating multi-scale ResNet models with deep feature fusion. Brief. Bioinform. 23: bbab530.

[8] Katsamenis, I., Protopapadakis, E., Voulodimos, A., Doulamis, A. and Doulamis, N. (2020). Transfer learning for COVID-19 pneumonia detection and classification in chest X-ray images. pp. 170–174. *In*: 24th Pan-Hellenic Conference on Informatics.

[9] Chen, J., Ma, Q. and Wang, W. (2021b). A lung cancer detection system based on convolutional neural networks and natural language processing. pp. 354–359. *In*: 2021 2nd International Seminar on Artificial Intelligence, Networking and Information Technology. AINIT.

[10] Lyu, Lei. (2021). Lung cancer diagnosis based on convolutional neural networks ensemble model. *In*: 2021 2nd International Seminar on Artificial Intelligence, Networking and Information Technology. AINIT. IEEE.

[11] Asuntha, A., Srinivasan, A.J.M.T. and Applications. (2020). Deep learning for lung cancer detection and classification. 79: 7731–7762.

[12] Tan, M. and Le, Q. (2019). EfficientNet: Rethinking model scaling for convolutional neural networks. pp. 6105–6114. *In*: International Conference on Machine Learning.

[13] Narin, D. and Onur, T.O. (2022). The effect of hyperparameters on the classification of lung cancer images using deep learning methods. Erzincan Univ. J. Sci. Technol. 15: 258–268.

[14] Kareem, H.F., AL-Husieny, M.S., Mohsen, F.Y., Khalil, E.A. and Hassan, Z.S. (2021). Evaluation of SVM performance in the detection of lung cancer in marked CT scan dataset. Indonesian J. Electr. Eng. Comput. Sci. 21: 1731.

[15] Raza, R., Zulfiqar, F., Tariq, S., Anwar, G.B., Sargano, A.B. and Habib, Z. (2022). Melanoma classification from dermoscopy images using ensemble of convolutional neural networks. Mathematics 10: 26.

[16] Humayun, Mamoona, Sujatha, R., Almuayqil, Saleh Naif and Jhanjhi, N.Z. (2022a). A transfer learning approach with a convolutional neural network for the classification of lung carcinoma. Healthcare 10(6): 1058. MDPI.

[17] Al-Yasriy, H.F., AL-Husieny, M.S., Mohsen, F.Y., Khalil, E.A. and Hassan, Z.S. (2020b). Diagnosis of lung cancer based on CT scans using CNN. IOP Conf. Ser.: Mater. Sci. Eng. 022035.

[18] Anon. (2023a). Chest CT-Scan images Dataset. Available: Chest CT-scan images dataset. https://www.kaggle.com/datasets/mohamedhanyyy/chest-ctscan-images. (Accessed 07 June 2023).

[19] Anon. (2023b). The IQ-OTH/NCCD lung cancer dataset — Kaggle. https://www.kaggle. com/datasets/hamdallak/the-iqothnccd-lung-cancer-dataset. (Accessed 19 January 2023).

[20] Jaszcz, A., Połap, D. and Damaˇseviˇcius, R. (2022). Lung X-Ray image segmentatioň using heuristic red fox optimization algorithm. Sci. Program. 2022: 1–8.

[21] Alyasriy, H. and Muayed, A. (2020). The IQ-OTHNCCD lung cancer dataset. Mendeley Data 1: 1–13 (Accessed 19 January 2023).

[22] Rehan Raza, Fatima Zulfiqar, Muhammad Owais Khan, Muhammad Arif, Atif Alvi, Muhammad Aksam Iftikhar et al. (2023). Lung-EffNet: Lung cancer classification using EfficientNet from CT-scan images. Engineering Applications of Artificial Intelligence 126(Part B): 106902, ISSN 0952-1976, https://doi.org/10.1016/j.engappai.2023.106902.

[23] Chen, Jiahao, Ma, Qianli and Wang, Weixin. (2021a). A lung cancer detection system based on convolutional neural networks and natural language processing. *In*: 2021 2nd International Seminar on Artificial Intelligence, Networking and Information Technology. AINIT. IEEE.

[24] Narin, D. and Onur, T.O. (2022). The effect of hyper parameters on the classification of lung cancer images using deep learning methods. Erzincan Univ. J. Sci. Technol. 15: 258–268.

[25] Kavitha, P. and Prabakaran, S. (2019). A novel hybrid segmentation method with particle swarm optimization and fuzzy c-mean based on partitioning the image for detecting lung cancer.

[26] Humayun, Mamoona, Sujatha, R., Almuayqil, Saleh Naif and Jhanjhi, N.Z. (2022a). A transfer learning approach with a convolutional neural network for the classification of lung carcinoma. Healthcare 10(6): 1058. MDPI.

[27] Azween Abdullah, Nithya, S., Mary Shanthi Rani, M., Vijayalakshmi, S. and Balamurugan Balusamy. (2023). Stacked LSTM and Kernel-PCA-based ensemble learning for cardiac arrhythmia classification. International Journal of Advanced Computer Science and Applications (IJACSA) 14(9): 2023. http://dx.doi.org/10.14569/IJACSA.2023.0140905.

[28] Vijayalakshmi, S., Sivakumar, V., Nataraj, C. and Kanth, P.C. (2023). Prediction of heart disease using XGB classifier, AIP Conference Proceedings Volume 3161, Issue 130 August 2024 Article number 020033, 5th International Conference on Sustainable Innovation in Engineering and Technology SIET 2023Kuala Lumpur 16 August 2023 Code 202231.

[29] Tanwar, S., Sivakumar, V., Nataraj, C. and Vijaylakshmi, S. (2023). Detection of colorectal cancer using dilated convolutional network via Raman spectra, AIP Conference Proceedings Volume 3161, Issue 130 August 2024 Article number 020030, 5th International Conference on Sustainable Innovation in Engineering and Technology 2023, SIET 2023 Kuala Lumpur 16 August 2023 Code 202231.

Deep Learning in Sentiment and Text Analysis

Chapter 5

Deep Learning Methods for Aspect Oriented Sentimental Analysis on Social Media Data

Samson Ebenezar Uthirapathy[1,*] and *Domnic Sandanam*[2]

1. Introduction

In the digital era, most individuals spend much of their time on interactive multimedia platforms such as Twitter, Facebook, Instagram, and others to express their thoughts, views, information, emotions, feelings, and verdicts. The convictions of products, events, organizations, sports incidents, and activities are very corroborative of publicization and initiatives. It is integral to evaluate the sentiments or convictions proliferated through interactive multimedia [1]. Categorizing people's beliefs into positive, negative, or neutral polarity is a process in sentiment analysis, also known as opinion mining. Aspect-based sentiment analysis (ABSA), also known as the attribute-based sentiment strategy, is used to discover various features, attributes, and aspects of a specified text and their respective sentiments. Utilizing customer feedback resources, such as reviews, survey results, and job opportunities in the healthcare industry that move from public relations to client indulgence to dispassionate prescribing, is highly practical.

ABSA particularly analyses aspects such as mobile phone reviews, feedback on medicines, treatment reviews, and drug reviews [2–4]. Here, tweets about medicines for various diseases have been considered, along with the opinions of common people and experts about the outcome of the medicines [5]. Sentiment analysis works on public convictions and is helpful in the decision-making process. This decision

[1] Department of Computer Science, Saveetha School of Engineering, SIMATS, Chennai, India.
[2] Department of Computer Applications, National Institute of Technology, Tiruchirappalli, India.
* Corresponding author: u.samson@gmail.com

DOI: 10.1201/9781003635178-7

will enhance the organization's growth and satisfy customers. The text consists of unstructured data that needs to be processed by eliminating unwanted strings [6]. Currently, through interactive multimedia, specifically Twitter, common mankind, experts, and physicians have debated various health-concerning issues. These debates are in the form of reviews of medicines or drugs prescribed for various illnesses.

The data retrieved from Twitter in the form of tweets is properly analysed to get insights into the discussion about medicines and the patients' conditions when using a particular drug [7]. The classification of the feedback that the patient gives will be helpful for the organization to understand the patient's condition after the use of the drug. The results of the labels will help make decisions on the drug to be deployed on the market. For example, if the price of the drug is high but the service provided by the organization is good and hygienic, the verdict assigns positive polarity to the aspect service and negative polarity to the aspect price in one verdict. Sometimes the clients can scribble reviews of the drug on multimedia about how the drug is working effectively or ineffectively. If there are numerous aspects to the reviews, correctly predicting the sentimental polarity becomes a challenge in sentimental analysis. The ABSA finds aspect-level polarity in the opinions in the reviews. It consists of two steps [8–10].

One feature is aspect-term identification and sentiment analysis. In these aspects, numerous explorations customarily make use of machine learning strategies for sentimental analysis in a supervised manner nowadays. RNN and CNN are two popular deep-learning methods used extensively in sentiment analysis. Deep learning strategies can comprehend convoluted pictures, text, sounds, and other data patterns to enable accurate perception and prediction.

Usually, for the sentimental analysis, there are primarily five steps as shown in Figure 5.1: preparation of data, evaluation of reviews, classification of sentiments,

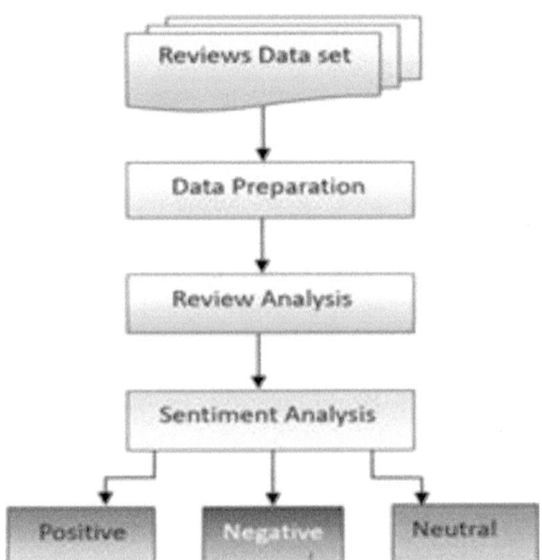

Figure 5.1 Steps of sentimental analysis.

and their outcomes that transfer the reviews into three different categories, namely positive, negative, and neutral.

This research article proposes a strategy for improving the accuracy level of ASBA classification with the modified RNN method, which is a combination of LSTM and GCN. The important contribution of the proposed model is as follows: adopting the finest feature extraction methodology to improve the model's classification performance. Scheming of a view classification method which uses the LSTM with GCN technique for improving the accuracy of the classification model.

2. Related Work

In this portion, some of the recent and past works based on our proposed model are being discussed.

Soni and Mathur [11] have planned an LSTM with an encoder attention mechanism for sentiment analysis. Here, they used attention mechanisms to extract the aspect features and then went through an LSTM for sentiment classification. Wadawadagi and Pagi [12] have offered a novel method that comprises Bidirectional LSTM with conditional random field (CRF) and a Polarity Enriched Attention Neural (PEAN) model to extract aspect terms and categorization of sentiments.

An innovative method for quantifying sentiment-oriented classification has been anticipated by Gurunathan [13]. The work aimed to fine-tune the sentiment orientation detection task by appropriately weighting the modifiers' or intensifiers' intentions based on the level of intensification and by accounting for important factors such as adjectives, negations, double intensification, and numerically expressed sentiments. Divate [14] examined the sentiments contemporary in the Marathi tweets using LSTM.

Uthirapathy and Sandanam [15] anticipated a novel technique using fuzzy c means, confined to sentiment classification in tweets. Ugochi and Prasad [16] established a classification framework for sentiment analysis. The authors utilised Naive Bayes for predicting the classes. This was compared against other methods like SVM and KNN. A hierarchical model using LSTM for aspect-oriented sentiment analysis was deliberated by Ruder et al. [17].

Liang et al. [18] proposed a method to recognise and extract feature words and estimation words using a dependency tree-based method called Senticnet, along with a GCN. The exclusive affective enhanced graph model, which consists of the contextual word-aspect and word dependencies, along with the opinion-aspect word emotional evidence. To guarantee that the SemGCN module correctly captures semantic relationships between words, Li et al. [19] recommended making use of orthogonal and differential regularizations to provide limits on courtesy ratings. The orthogonal standardisation aids the SemGCN to find more strictly connected idioms by reducing the amount of intersection between the words being learned. Since the SynGCN may have missed some semantic signals, the differential regularizer encourages the SemGCN to fill in the gaps.

Karimi et al. [20] offered an original approach based on Adversarial Training using BERT by applying adversarial learning to the two key goals in sentiment analysis: Aspect term extraction and Sentiment categorisation (BAT). Hu et al.

[21] proposed a Multi-level Semantic Relation-enhanced Learning Network (MSRL-Net), a method based on ABSA, as well as a self-supervised task to classify pairs of expressions based on the semantic associations. For the sentence semantic matching job, MSRL-word-level Net uses word-dependency relations and sentence-word relationships in the sentence to augment its semantic depiction; for the sentence pair relation cataloguing test, MSRL-sentence-level Net uses the semantics of sentence relations and the relations of sentence pairs.

3. Proposed Methodology

The four basic components of the anticipated methodology are discussed in this section, which consist of the classification of sentiments and their aspect features. As shown in Figure 5.2, the proposed model consists of data collection, data preprocessing, which involves pattern removal, tokenization, and stemming, then the features are extracted, aspect extraction, and finally the aspect is classified into three categories: positive, negative, or neutral.

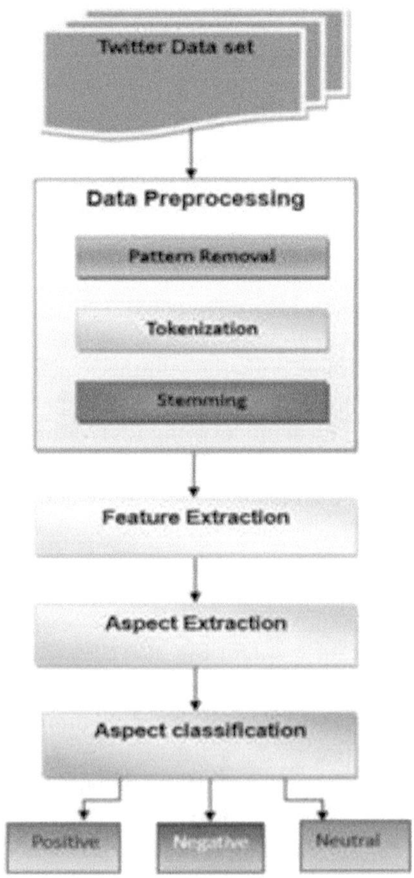

Figure 5.2 Proposed Method architecture.

3.1 Data Pre-Processing

Data preprocessing needs to be done for all the words, images, and also for the special symbols that are available in the reviews that are included in the interest of the study. It also involves the removal of irrelevant single quotes, email addresses, and newline characters, and also tries to remove the contextual noise of the data. The complete text has been converted to lowercase, and we also removed all punctuation in consideration of the following review.

After using Fingolimod, there is severe bleeding in the right eye. I have been in pain for thirty minutes. The punctuation and numerals are eliminated because the review contains the time 30, which is both a number and a period, as shown in Figure 5.3. As sentiment analysis is not affected by the mentioned elements, they are left in. Moreover, many stop words that are generally defined as words similar to "is off," "of," etc., are removed.

Stemming is a method of withdrawing affixes to convert an expression into its root form. Many words in the reviews are not in their root forms. This process is known as lemmatization. Furthermore, the processed dataset is used to extract attributes in the next pattern.

Tokenization plays a significant role in pre-processing. Here, the sentences are divided into individual words.

	text	sentiment
0	An adverse drug reaction (ADR) is an injury ca...	neutral
1	unpredictability issues have been occured with...	negative
2	An infectious disease caused by protozoan para...	negative
3	An allergy is an immune system response to a f...	negative
4	Hypothyroidism, also called underactive thyroi...	negative

Figure 5.3 Data Pre-processing.

3.2 Word Embeddings

After data pre-processing is completed, the dataset has undergone a word embedding process. It converts words into vectors. Here, we considered two, namely the TF-IDF method and the BERT method. So, two different input data sets and the performance comparison between the two variations are explained in the results and discussion section.

3.2.1 Word Embedding Using TF-IDF

It refers to the occurrence of terms T that occur in Document D. The IDF is a measurement of whether a term is frequent or not in a distinct document.

$$tf - idf(t, d) = tf(t, d) * idf(t) \quad (1)$$

In equation (1) *tf*: term prevalence, *idf*: adverse demonstrate regularity, t: a term present in the document; *d*: a document that contains the frequency of the term will be attained by multiplying the TF terms (*t*) and documents with the IDF terms (*t*, *d*). It is a fashionable encryption strategy for managing categorical attributes. In this strategy, each attribute is imposed as a personalized integer based on alphabetical ordering. The words are converted to numbers so that computers can understand and process them.

3.2.2 Word Embedding Using BERT

Natural language processing can utilise the potential of the deep learning process known as Bidirectional Encoder Representation from Transformers, called BERT. It is unique among language models in that it can read input sequences in both the traditional left-to-right and right-to-left formats. The Transformer fuels the BERT, which uses an attention mechanism to investigate proper word pairings in a phrase. A Transformer typically has an encoder built in. The input texts are read by the encoder. In this study, we use BERT [22] to solve the problem of classifying people's emotions.

A token sequence is an input to the encoder. Before being sent to the encoder, the inputs are transformed into vectors known as word embeddings. Here, we use word embedding as a hybrid of token, segment, and position embeddings. Two special tokens are embedded in the input word tokens by a token embedding technique. Classification tokens [CLS] and [SEP] are placed before and after the first word of the sentence, respectively. [SEP] can tell when one sentence ends, and another begins within the same input sequence.

In segment embeddings, a marker is appended to each token to help the encoder distinguish between phrases. To identify its location in the text, a positional embedding is applied to each token.

3.3 Combination of LSTM and GCN in Classification Layer

The steps that are involved in combining the LSTM and GCN as an embedded input in the proposed model in the classification layer are discussed in this section.

3.3.1 Inputs to the Model

Consider the situation where $c = \{c_1, c_2, c_3, \ldots, c_n\}$ and also we can take the aspect term as *e*, the comparative distance is s_j which is measured amongst the jth context word and that particular aspect term which has been included in an axiom. If any specific feature term has numerous diverse units, then the average of the total word embeddings of diverse units will fix the insertion value of the target aspect term. The time engaged for passing is represented by the numeral *n*. LSTM+GCN considers the overall environment along with the particular component that is targeted as the inputs.

3.3.2 Aspect Term Embedding and Context Term

The one-hot vector format needs to be converted into a word embedding earlier to nourish the context and aspect target into layers of LSTM+GCN. The given term is changed from a vector having higher dimensions and discrete points to a vector having a smaller size and focused points. By executing a mathematical function in the embedded space, the compositionality can be computed. Given the matrix $M \in R^{V \times d}$ of the word entrenching, where the letter V is used to represent the vocabulary size, and the letter d represents the entrenching dimension of the word. A dependency matrix example is shown below, which is used for acquiring the context embedding c_j' with the lookup table utilising equation (2).

$$c_j' = lookup(c_j) = c_j^T \cdot M \tag{2}$$

From the term embedding matrix M, the aspect embedding e_j' can be attained using equation (3).

$$e_j' = \frac{1}{r}\sum_j^r lookup(e_j) = \frac{1}{r}\sum_j^r e_j^T \cdot M \tag{3}$$

3.3.3 Location Embedding of the Word

The position embedding of the mined context word along with the aspect term is calculated by using the relative distance between the context word and the specific aspect term created from equation (4) if the aspect term contains one word.

$$s_j = |i^s - j^s| \tag{4}$$

Here i^s represents a particular aspect term in the sentence and j^s represent the context word in the sentence.

If the aspect term consists of multiple words e $=\{e_1, e_2, e_3, ..., e_n\}$. The relative distance can be calculated as mentioned in equation (5).

$$s_j = \{i - j,\ \textit{if context word } c_j \textit{ is left of } e_j - (i + r),\ \textit{if context word } c_j \textit{ is right of } e \tag{5}$$

The location embedding l_j is then computed with the lookup table using equation (6).

$$l_j' = lookup(l_j) = l_j^T \cdot M \tag{6}$$

By concatenating each word using location embedding, context embedding, and aspect term embedding, the final input vector is created.

3.4 LSTM Layers

In the following step, the participation vector is fed into the LSTM layer to make it study the semantics of the sentences by processing each word in sequence, as given in Figure 5.4. The LSTM layers extend the short-term memory, which comprises cells. Every cell holds an output gate, a forget gate, and one or more input gates. The first one is the forget gate, which acts as the filter and allows the relevant information [23] and discards the irrelevant information by using the sigmoid stimulation function. The sigmoid activation function maintains the vectors between [0, 1]; if the value is close to 0, it will be discarded, and the values close to 1 will be allowed. Input gates select which portions of new data to maintain in the existing state using the same process as forgetting gates.

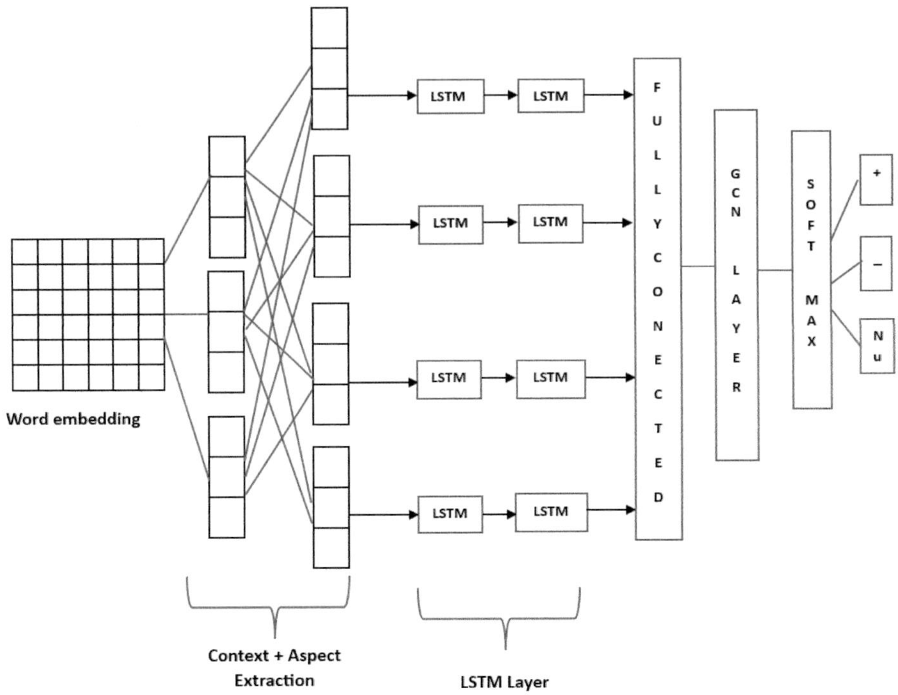

Figure 5.4 LSTM+GCN Model.

The LSTM network's input gate is mathematically expressed in equation (7).

$$i_t = \sigma(w_i\,[h_{t-1}, x_t] + b_i) \tag{7}$$

The LSTM network's forget gate is mathematically expressed in equation (8).

$$f_t = \sigma(w_f\,[h_{t-1}, x_t] + b_f) \tag{8}$$

$$o_t = \sigma(w_o\,[h_{t-1}, x_t] + b_o) \tag{9}$$

In Equations (8) & (9) f_t represent the overlook gate value, σ is the sigmoid initiation function, w_f weight for the forget gate, h_{t-1} is the output of the preceding layer of the LSTM at timestamp $(t-1)$, x_t input at the current timestamp, b_f, b_o biases for the forget and output gates, respectively.

The cellular gate formula for the new evidence, as mentioned in equation (10)

$$S_t' = tanh(w_S\,[h_{t-1}, x_t] + l_S) \tag{10}$$

In this instance, S_t is a candidate of the cell state during timestamp t, w_S is a weight for the neurons in the appropriate cell state S, and l_S is a bias for the appropriate cell state S.

The updated cell state at time stamp t is mentioned in equation (11).

$$S_t = f_t * S_{t-1} + i_t * S'_t \tag{11}$$

Where, S_t Candidate of cell state with timestamp t, R_t Output cell at timestamp t, P_t Input cell at timestamp t, and S'_t Cell state during the timestamp.

The output of the LSTM hidden layer is given by the equation below (12).

$$h_t = o_t * tanh\ tanh\ (S^t) \tag{12}$$

3.5 Graph Convolutional Networks

The value of h_t is sent to the GCN layer and then to the final softmax output layer. The GCN layer takes the dependency matrix as input, and the GCN deliberates the sentence and computes the dependency between the words using a convolutional graph [18]. The new dependency tree G with k nodes is constructed, where a sentence S having N words is considered. Here, the words are taken as the nodes, and the relationship between any two words is considered as the edges. Hence, it is likely to rebuild the dependency matrix with the LSTM layer output as $X_{ij} \in R^{k*k}$; 1 is used for the dependent position, and 0 is used for the other situations. h_i^{l-1} characterises the input to the ith node, while h_i^l denotes the output. Equations (13) and (14) are the calculations for the graph convolutional neural network.

$$h_i^l = \sigma\left(\sum_{j=1}^{n} X_{ij} W^l h_i^{l-1} + b^l\right. \tag{13}$$

Here, W^l is the weight and b^l is bias the X_{ij} matrix will be replaced with a dependency matrix D_{ij} the formation of D_{ij} is represented in equation (14).

$$D_{ij} = X^{ij} * (L_{ij} + P_{ij} + \Delta) \tag{14}$$

Using equations (13) and (14), the depiction of the sentence in the last hidden layer is updated as mentioned in equation (15).

$$H^m = \sigma\left(\sum_{j=1}^{n} D_{ij} W^l h_i^{l-1} + b^l\right. \tag{15}$$

The H^m is the aspect features that are extracted by the model GCN, which is used to efficiently represent the aspect words in the sentences.

3.6 Output Layer

The aspect feature representations in sentence S are sent to the output layer, where the divergence is classified as positive, neutral, and negative. The output layer activation function is SoftMax, which is represented in Equation 16.

$$y(a) = softmax\ (w_o S + b_0) \tag{16}$$

3.7 Optimizer

The projected LSTM-GCN model uses the Adam backpropagation technique, which aids in fine-tuning the learning rate to increase the recommended model's accuracy

In mathematical representation, Adam is signified in equations 17 and 18.

$$w_{t+1} = w_t - \alpha m_t \tag{17}$$

$$m_t = \beta m_{t-1} + (1-\beta)\left[\frac{\delta L}{\delta w_t}\right] \tag{18}$$

In equations 17 and 18, m_t represent the combination of the gradients at time t and m_{t-1} is the gradient aggregation during time t–1. w_t, w_{t+1} Are the weights at time t and t–1, respectively? α is learning, and β is the moving average parameter, which will take the constant value of 0.9. δL is the loss function derivative at time t and δw_t is the weight derivative.

4. Results and Discussion

The recommended method LSTM+GCN, is assessed with the two openly accessible datasets available in Kaggle website.

4.1 Dataset Description

The proposed work uses two data sets to find the efficacy of the given LSTM+GCN technique for aspect-based sentiment categorization. The data sets have reviews connected with the drugs.

4.1.1 Health-Related Data on Drugs

The Healthcare Related Dataset was acquired from [28]. The dataset includes "participant reviews who take the medication." Text ID, text reviews, labels, and the type of medicine taken are among the four separate columns included in the dataset. It consists of 4 columns and 27503 rows. The reviews will be divided into favourable, negative, and neutral categories.

There are 27,481 occurrences in the training dataset, which has three categories: 8,582 rows containing positive reviews, 7,781 rows containing negative reviews, and 11,118 rows containing neutral reviews. The testing dataset consists of 3534 cases, with three labels: 1,103 positive, 1,001 negative, and 1,430 neutral.

4.1.2 UCI ML Drug Review

The dataset UCI ML comprises patient reviews on explicit drugs along with the required situations. The rating system in the data set follows a 10-star rating, which reflects the overall patient satisfaction. The reviews, along with the sentiment of various illnesses, are included in the collection [24]. Here, ratings 1 to 3 are considered negative ratings. The negative ratings are 35064 rows in the training data set and 11,839 in the test dataset, while the ratings 4 to 6 are considered neutral, in the training data set 19,369 rows and 6,489 rows in the test data set. The ratings 7 to

10 are considered positive ratings. There are 106,867 rows in the training data set and 35,441 rows in the test data set.

The model was assessed with statistics. In accordance with the guidelines, precision, recall, and F-score were calculated.

4.2 Performance Evolution

To boost performance, ensure stability, and increase accuracy, many parameters were used as mentioned in Table 5.1.

Table 5.1 Settings of Parameters.

Parameters	**Values**
Layers of LSTM	3
Layer of GCN	1
Softmax Layer	1
Optimizer (Backpropagation)	RMSprof
Word embeddings	TF-IDF, BERT
Learning rate TF-IDF	1e -3
Learning rate BERT	2e -4
Dropout	0.5
Batch Size	128
Number of Epochs	100

Precision

Precision in the performance measures the proportion between the number of true positive values in the data and the number of predicted positive values by the classifier. It is mentioned in equation 19.

$$Precision = \frac{TP}{TP + FP} \tag{19}$$

Recall

Recall means the proportion of the true positive data present in the dataset, with all the positive observations in the class. A recall is calculated using equation 20. A high recall value means a low false rate.

$$Recall = \frac{TP}{TP + FN} \tag{20}$$

F1-Score is mentioned in equation 21.

$$F - Score = 2 * \frac{(Precision * Recall)}{(Precision + Recall)} \tag{21}$$

Accuracy

Accuracy is the measure that provides equal importance to all classes. Accuracy is calculated using the equation 22.

$$Acc' = \frac{TP + TN}{TP + TN - FP + FN} \tag{22}$$

Whereas TP stands for the proportion of positive reviews categorized as positive, TN for the proportion of negative values of reviews categorized as negative, FP for the classification of positive reviews with negative review sentences, and FN for the classification of positive reviews incorrectly identified as negative.

With the aid of publicly accessible data, the performance was assessed with four different approaches (LSTM, BI-LSTM, GRU, and simple RNN).

The experiments explain the performance of various algorithms. The values from Tables 5.2 to 5.5 show the results of various embedding models. From the table, we can see that the anticipated model, LSTM with GCN, is showing better results when compared with other classification models.

Table 5.2 Classification report on Health data on drugs with the TF-IDF word embedding method.

Methods	**TF-IDF Word Embedding**			
	Precision	**Recall**	**F-Score**	**Accuracy**
LSTM	0.89	0.87	0.88	0.88
LSTM+GCN	0.92	0.91	0.92	0.93
Bi-LSTM	0.82	0.92	0.85	0.87
GRU	0.89	0.83	0.86	0.85

Table 5.3 Classification report of health-related data on drugs using the BERT word embedding method.

Methods	**BERT Word Embedding**			
	Precision	**Recall**	**F1-Score**	**Accuracy**
LSTM	0.92	0.91	0.91	0.91
LSTM+GCN	0.95	0.96	0.97	0.97
Bi-LSTM	0.85	0.93	0.88	0.90
GRU	0.91	0.85	0.89	0.88

Table 5.4 UCI ML drug review dataset classification report using TF-IDF word embedding method.

Methods	**TF-IDF word embedding**			
	Precision	**Recall**	**F-Score**	**Accuracy**
LSTM	0.86	0.85	0.87	0.85
LSTM+GCN	0.92	0.91	0.90	0.92
Bi-LSTM	0.87	0.82	0.85	0.87
GRU	0.85	0.86	0.88	0.83

Table 5.5 Classification report UCI ML drug review dataset with the BERT word embedding method.

Methods	BERT Word Embedding			
	Precision	Recall	F-Score	Accuracy
LSTM	0.89	0.87	0.88	0.89
LSTM+GCN	0.94	0.93	0.93	0.94
Bi-LSTM	0.83	0.82	0.87	0.88
GRU	0.89	0.88	0.90	0.87

5. Conclusion

The current research proposes an LSTM+GCN model, and it also proves the efficiency of the model when it is applied for sentiment analysis. The proposed model was experimented on two different data sets, along with two word embedding approaches, such as TF-IDF and BERT, which are used in the system. The proposed model has an LSTM layer, which was employed in the research for a feature extraction stage, followed by the GCN layer, which is for the classification of sentiment. We also employ a SoftMax layer that is used for standardising the multiclass output. The output of the experiments categorizes the participants' review of the sentences as good, neutral, or negative. The LSTM+GCN model outputs improved results when compared with those of BERT word embedding and TF-IDF word embedding. The anticipated model was related to other existing models for sentiment classification, where the results revealed that the proposed LSTM+GCN outperformed other models with 4% of precision and recall and 5% of accuracy in terms of performance.

References

[1] Li, J., Fong, S., Zhuang, Y. and Khoury, R. (2019). Hierarchical classification in text mining for sentiment analysis. pp. 46– 51. In 2019 International Conference on Soft Computing and Machine-Intelligence, IEEE.

[2] Parveen, H. and Pandey, S. (2018). Sentiment analysis on twitter data-set using naive bayes algorithm. pp. 416–419. In 2016 2nd International Conference on Applied and Theoretical Computing and Communication Technology (iCATccT), IEEE.

[3] Soumya, S. and Pramod, K. (2020). Sentiment analysis of Malayalam tweets using machine learning techniques. ICT Express 6(4): 300–305.

[4] Uthirapathy, S.E. and Domnic, S. (2020). Real-time opinion prediction method for emergency public events in social media networks using opinion hit matrix. Rev. d'Intelligence Artif. 34(4): 507–514.

[5] Tuhin, R.A., Paul, B.K., Nawrine, F., Akter, M. and Das, A.K. (2019). An automated system of sentiment analysis from bangla text using supervised learning techniques. pp. 360–364. In 2019 IEEE 4th International Conference on Computer and Communication Systems (ICCCS), IEEE.

[6] Devika, M., Sunitha, C. and Ganesh, A. (2020). Sentiment analysis: a comparative study on different approaches. Procedia Computer Science 87: 44–49.

[7] Cambria, E., Das, D., Bandyopadhyay, S., Feraco, A. et al. (2019). A practical guide to sentiment analysis.

[8] Medhat, W., Hassan, A. and Korashy, H. (2018). Sentiment analysis algorithms and applications: A survey. Ain Shams Engineering Journal 5(4): 1093–1113.

[9] Mejova, Y. (2020). Sentiment Analysis: An Overview. University of Iowa, Computer Science Department.

[10] Zhang, L., Wang, S. and Liu, B. (2018). Deep learning for sentiment analysis: A survey. Wiley Interdisciplinary Reviews: Data Mining and Knowledge Discovery 8(4): e1253.
[11] Soni, J. and Mathur, K. (2022). Sentiment analysis based on aspect and context fusion using attention encoder with LSTM. Int. J. Inf. Tecnol. 14: 3611–3618. https://doi.org/10.1007/s41870-022-00966-.
[12] Wadawadagi, R. and Pagi, V. (2022). Polarity enriched attention network for aspect-based sentiment analysis. International Journal of Information Technology 14(6): 2767–2778.
[13] Gurunathan, P. (2022). Joint aspect-opinion extraction and sentiment orientation detection in university reviews. International Journal of Information Technology 14(6): 3213–3225.
[14] Divate, M.S. (2021). Sentiment analysis of Marathi news using LSTM. International Journal of Information Technology 13(5): 2069–2074.
[15] Uthirapathy, S.E. and Sandanam, D. (2023). Predicting opinion evolution based on information diffusion in social networks using a hybrid fuzzy based approach. International Journal of Information Technology 15(1): 87–100.
[16] Ugochi, O., Prasad, R., Odu, N., Ogidiaka, E. and Ibrahim, B.H. (2022). Customer opinion mining in electricity distribution company using twitter topic modeling and logistic regression. International Journal of Information Technology 14(4): 2005–2012.
[17] Ruder, S., Ghaffari, P. and Breslin, J.G. (2019). A hierarchical model of reviews for aspect-based sentiment analysis. arXiv preprint arXiv:1609.02745.
[18] Liang, B., Su, H., Gui, L., Cambria, E. and Xu, R. (2022). Aspect-based sentiment analysis via affective knowledge enhanced graph convolutional networks. Knowledge-Based Systems 235: 107643.
[19] Li, R., Chen, H., Feng, F., Ma, Z., Wang, X. and Hovy, E. (2021, August). Dual graph convolutional networks for aspect-based sentiment analysis. pp. 6319–6329. In Proceedings of the 59th Annual Meeting of the Association for Computational Linguistics and the 11th International Joint Conference on Natural Language Processing (Volume 1: Long Papers).
[20] Karimi, A., Rossi, L. and Prati, A. (2021, January). Adversarial training for aspect-based sentiment analysis with bert. pp. 8797–8803. In 2020 25th International Conference on Pattern Recognition (ICPR). IEEE.
[21] Hu, Z., Wang, Z., Wang, Y. and Tan, A.H. (2023). MSRL-Net: A multi-level semantic relation-enhanced learning network for aspect-based sentiment analysis. Expert Systems with Applications p.119492.
[22] Xu, H., Liu, B., Shu, L. and Yu, P.S. (2019). Bert post-training for review reading comprehension and aspect-based sentiment analysis. arXiv preprint arXiv:1904.02232.
[23] Huang, B., Guo, R., Zhu, Y., Fang, Z., Zeng, G., Liu, J. et al. (2022). Aspect-level sentiment analysis with aspect-specific context position information. Knowledge-Based Systems 243: 108473.
[24] https://www.kaggle.com/datasets/jessicali9530/kuc-hackathon-winter-2018?select=drugsComTest_raw.csv.
[25] Kamalesh, B. and Vijayalakshmi, S. (2022). Sentiment analysis on amazon product review. pp. 846–848. 2022 4th International Conference on Advances in Computing, Communication Control and Networking (ICAC3N), Greater Noida, India, doi: 10.1109/ICAC3N56670.2022.10074287.
[26] Ramesh, R. and Vijayalakshmi, S. (2022). Improvement to Recommendation system using Hybrid techniques. pp. 778–782. 2022 2nd International Conference on Advance Computing and Innovative Technologies in Engineering (ICACITE), Greater Noida, India, doi: 10.1109/ICACITE53722.2022.9823879.
[27] Yadav, K. and Vijayalakshmi, S. (2022). Social media sentiment analysis of stock market on tweets. pp. 1977–1980. 2022 2nd International Conference on Advance Computing and Innovative Technologies in Engineering (ICACITE), Greater Noida, India, doi: 10.1109/ICACITE53722.2022.9823529.
[28] https://www.kaggle.com/datasets/arpikr/uci-drug.

Chapter 6

Algorithmic Strategies for Copyright Protection in Digital Streaming Services through Geo-blocking and Geo-filtering

Deepika Chhangani,[1] *Avishek Chakraborty*[2] and *Sanjay Bang*[3,*]

1. Introduction

As global streaming subscriptions have reached two billion users by the end of 2024, safeguarding digital content rights has become a critical challenge for Over-the-Top (OTT) platforms [1]. OTT services like Netflix, Disney+, Amazon Prime Video, and other providers have reshaped media consumption, offering on-demand access across borders and devices. The borderless, widespread distribution of digital content clashes with traditional territorial licensing models, creating challenges for enforcing copyright protection and censoring without affecting the user experience.

Geo-blocking is a technique that restricts access to content based on IP address, billing information, or contractual territory, which serves as the primary technical barrier against unauthorised streaming [2]. The increased use of Virtual Private Networks (VPNs) and anonymisation tools has rendered static IP-based filters increasingly porous: empirical studies show that many commercial VPN services routinely bypass geo-blocking countermeasures, enabling unauthorised access

[1] Research Scholar, Christ University, India.
[2] School of Law, Christ University, India.
[3] Associate Professor, Christ University, India.
Emails: deepika.chhangani@res.christuniversity.in, avishek.chakraborty@christuniversity.in
* Corresponding author: sanjay.satyanarayan@christuniversity.in

DOI: 10.1201/9781003635178-8

to content and revenue leakage [3]. Implementing rigid blocking harms the user experience, frustrates the legitimate subscribers, and undermines monetisation goals, as licensing agreements and jurisdictional claims grow ever more complex.

We analyse the current techniques used to enforce copyright protection in digital streaming services through geo-blocking and geo-filtering techniques, and propose a hybrid algorithmic–legal framework that transcends static geo-blocking by integrating dynamic, policy-driven geo-filtering with embedded legal compliance checks. Applying the algorithmic framework helps in the filtering of content in a geographical location base on the legal guidelines in that area, the key merits of this technique includes the following: (1) the formalization of geo-filtering as a smooth scoring function that adapts access levels per session and content item; (2) the encoding of key legal doctrines such as safe-harbour limits and intermediary-liability thresholds into real-time audit triggers; and (3) the design of a modular architecture offering transparent auditing and extensibility for evolving regional regulations. This approach balances user access, revenue protection, and legal compliance, paving the way for a constitutionally sound, self-regulatory code in digital streaming services.

2. OTT Platforms and Copyright

Significant limitations in conventional geo-blocking motivate the shift to geo-filtering; it is essential to understand the rapid rise of OTT platforms and the underlying copyright models they must navigate. The explosive growth of OTT services such as Netflix, Disney+, Amazon Prime Video, and others serves over 5 billion global users and generates hundreds of billions in annual revenue [6, 7]. We also examine how traditional territorial licensing works with exclusive rights per country or region, and how it interacts with Internet Protocol (IP) spoofing and globally distributed content delivery networks [8, 9]. We also discuss the practical shortcomings of geo-blocking, including widespread VPN circumvention, subscriber friction during travel, and the unintended consequences of high-profile enforcement actions like Netflix's 2023 password-sharing crackdown [10]. This background understanding will help clarify the need for more adaptive, policy-driven access controls and the need for geo-filtering mechanisms.

2.1 OTT Evolution

The past decade has seen explosive growth in OTT streaming services. In early 2025, global OTT users surpassed 5.27 billion [11], generating an estimated $316.4 billion in revenue, with Netflix alone accounting for 277 million paid subscribers and $33.7 billion in 2023 revenue [6]. Netflix continues to lead the market with over 301.6 million subscribers as of January 2025, followed by Disney+ and Amazon Prime Video [7]. Such a scale underscores the centrality of OTT platforms in modern media consumption and the economic stakes of securing digital rights.

2.2 Copyright Licensing Models

Traditional content licensing hinges on territorial rights, in which distributors acquire exclusive permissions to stream specific works within defined regions. Contracts specify which countries or markets a platform may serve, often tied to complex royalty and pricing schemes [14]. Under Regulation (EU) 2018/302, providers "may restrict or block access to online interfaces" outside licensed territories [8, 15, 17]. However, the digital environment erodes these clear boundaries because IP-based geolocation can be spoofed, and content delivered via global Content Delivery Networks (CDNs) can be delivered anywhere instantly. As a result, static territorial licenses face increasing tension between legal obligations and technical realities.

2.3 Limitations of Geo-Blocking

Geo-blocking is a technique used to deny access whenever a user's IP address falls outside a licensed region, which has been the primary enforcement tool. Yet, its rigidity and reliance on IP mapping introduce several shortcomings:

- *Circumvention via VPNs*: Millions of users exploit commercial VPN services to appear within permitted regions, bypassing geo-blocks and accessing unlicensed content. A 2015 study found over 30 million Netflix users employed VPNs or proxies [9, 16].
- *User Friction*: Legitimate subscribers travelling abroad or using mobile networks frequently face unnecessary denial of service, undermining customer satisfaction and brand loyalty.
- *Case in Point*—Password-Sharing Crackdown (2023): In May 2023, Netflix implemented a global password-sharing crackdown using the IP-based primary-location checks to enforce "one household" rules. Despite initial backlash, the measure boosted paid subscriptions, demonstrating the power of geo-based enforcement and consumer friction [10].

These limitations highlight the need for more adaptive, policy-driven mechanisms that reconcile licensing complexity, monetisation goals, and user experience, motivating the shift toward geo-filtering explored in this chapter.

3. Technical and Legal Review on Content Filtering

This section surveys prior work at the intersection of computer science and law, focusing on (1) technical enforcement mechanisms for territorial control, (2) statutory liability regimes, (3) industry self-regulation in practice, and (4) implications of algorithmic governance.

3.1 Technical Enforcement Mechanisms

We analysed the work of geo-blocking and geo-filtering. Traditional geo-blocking enforces territorial licenses by denying all access when a session's IP address falls outside a permitted region. While conceptually simple, this "hard" filter suffers from high false-positive and false-negative rates in practice [2, 12, 13]. Geo-filtering refines this approach by computing a policy score $P(u, c)$ that aggregates geolocation confidence, user credentials, and anonymiser detection, then comparing it against a content-specific threshold θc. Real-world deployments attest to both approaches: Netflix's 2023 password-sharing crackdown used primary-location geo-blocks to enforce "one household" rules, resulting in a 4% uptick in paid sign-ups and a 2% subscriber churn due to over-blocking. In contrast, a pilot geo-filtering system at a regional broadcaster in 2024 achieved 92% accuracy (TPR) with only 5% FPR, demonstrating the viability of soft-decision algorithms in live streaming environments [10, 18].

3.2 Legal Frameworks and Intermediary Liability

Safe-Harbour Protections. Under the U.S. Digital Millennium Copyright Act (DMCA), online service providers may avoid liability for user-generated infringements if they promptly remove infringing content upon notice [19]. Similarly, India's Information Technology Act shields intermediaries from third-party liability, provided they act as "mere conduits," do not initiate transmission, and respond to takedown requests [20]. OTT platforms distribute and often produce content, blurring the "conduit" distinction and leaving them vulnerable to contributory liability claims. Moreover, neither statute addresses proactive enforcement or circumvention via VPNs, exposing a gap between reactive takedown regimes and the need for real-time access control.

3.3 Self-Regulatory Codes in Practice

In response to statutory shortcomings, jurisdictions have turned to industry-led frameworks. India's Information Technology (Intermediary Guidelines and Digital Media Ethics Code) Rules, 2021, established a three-tier grievance redressal mechanism for OTT platforms:

- Self-regulation by the platforms' bodies, enforcing a Code of Ethics.
- Self-regulatory bodies of publishers oversee content classification and user complaints.
- Inter-departmental oversight, chaired by a government nominee, to adjudicate unresolved grievances.

Early reports indicate mixed results in addressing the grievance volumes, which declined 18% in late 2023 due to more straightforward content-classification guidelines. Platforms still face ambiguity when applying age-based filters to user-generated subtitled and dubbed content. This suggests self-regulation can standardise compliance but may struggle with the technical integration of policy and practice [21–23].

3.4 Algorithmic Governance and Discrimination

Algorithmic enforcement introduces risks of geodiscrimination, where legitimate users in high-risk regions face disproportionate access denial. A 2022 study found that geo-filtering systems trained on IP-region mappings misclassified 6% of users from small island nations as unauthorised, due to sparse training data and CDN anomalies. This bias can exacerbate digital divides and raise questions under principles of non-discrimination in international trade law [24, 25]. Mitigating such bias requires transparent auditing, periodic retraining with diverse samples, and fallback mechanisms featuring them as integral to our proposed audit-enabled framework.

4. A Hybrid AI Algorithmic-Legal Framework

We propose an effective architecture that unifies technical enforcement and legal compliance through content filtering, while maintaining the copyright of digital content within a single, modular system. By blending geo-filtering algorithms with embedded audit rules, this proposed framework enables OTT platforms to enforce territorial licenses dynamically while satisfying statutory obligations in real time.

4.1 Conceptual Model

The framework we propose comprises four core modules:

4.1.1 Geo-Data Collector

Gathers session metadata (IP address, GPS coordinates, CDN edge node ID) and user credentials (subscription tier and billing address).

4.1.2 Policy Engine

Computes access decisions via geo-blocking and geo-filtering functions and evaluates policy rules.

4.1.3 Anonymiser Detector

Applies heuristics and machine-learning classifiers to detect VPNs, proxies, and TOR usage, producing a penalty score v(u).

4.1.4 Audit Logger

Records every decision and triggers legal-audit flags when predetermined conditions are met, ensuring transparency and traceability.

Figure 6.1 shows the schematic diagram of the hybrid algorithmic-legal framework. The framework has four core modules and their interactions. The Geo-Data Collector gathers session metadata, including the IP address, GPS coordinates, CDN node information, and subscription credentials, and feeds these inputs into the Policy Engine, which also receives an anonymiser penalty score from the Anonymiser Detector (flagging VPNs, proxies, or TOR usage) [28]. The Policy Engine evaluates geo-blocking and geo-filtering decision functions to determine access levels and applies encoded legal audit rules simultaneously. The Audit Logger records all access decisions and the audit flags triggered, ensuring complete transparency and traceability. Feedback loops between the audit logger and geo-data

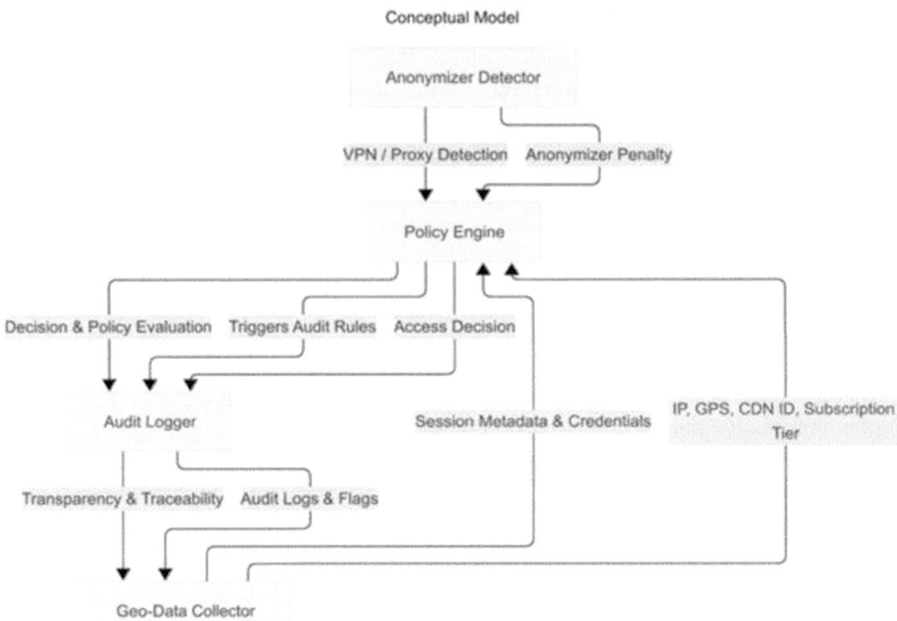

Figure 6.1 Schematic diagram of the hybrid algorithmic–legal framework.

collector enable ongoing refinement of policy weights and detection heuristics, supporting continuous improvement and compliance.

4.2 Formal Definitions

We define geo-blocking and geo-filtering as follows:

4.2.1 Geo-Blocking

$$B\,(u,\,c) = 1,\; if\, loc(u) \in R_c\; else\; 0 \tag{1}$$

Where u is the user session, c is the content item, and R_c is the set of licensed regions for c.

Legal linkage can be identified and denied if $B\,(u,\,c) = 0$, which enforces territorial license clauses under the DMCA and EU Regulations or any applicable regulations.

4.2.2 Geo-Filtering

We define the continuous decision function using a logarithmic transform:

$$F(u,c) = \sigma(P(u,c) - \theta_c), \sigma(x) = \frac{1}{1+e^{-x}} \tag{2}$$

With the policy score

$$P(u,\,c) = \omega_1\, g(loc(u),\,c) + \omega_2\, h(cred(u)) - \omega_3\, v(u) \tag{3}$$

where the components in equations (2) and (3) represent the following,

- $g(loc(u), c) \in \{0, 1\}$ indicates region match,
- $h(cred(u)) \in [0, 1]$ reflects subscription and billing validation,
- $v(u) \in [0, 1]$ is the anonymiser penalty from the Anonymiser detector.
- $\omega_1 + \omega_2 + \omega_3 = 1$ are tunable weights, and
- θ_c is a content-specific access threshold derived from licensing agreements.

The threshold θ_c encodes contractual restrictions (e.g., pay-per-view versus subscription rights) and intermediary liability obligations per various Information Technology (IT) Acts [26, 27, 29]. As per the framework, a session is granted full access if $F(u, c) \geq 0.5$; otherwise, access may be degraded and, in turn, denied.

4.3 Policy-Rule Encoding

Legal doctrines and self-regulatory code clauses are encoded as audit rules evaluated after each decision. The pseudocode represents the rules and the trigger for advanced audit, when the check is required using advanced parameters and usage patterns:

```
// Rule 1: "Mere Conduit" audit trigger
if isOriginalContent(c) and platformActsAsDistributor then
  AuditLogger.flag(u, c, "MereConduitCheck")
endif
// Rule 2: Safe-Harbour proactive audit
if v(u) > v_max and F(u,c) >= 0.5 then
  AuditLogger.flag(u, c, "HighAnonymizerRisk")
endif
```

Rule 1 ensures any content produced or co-produced by the platform beyond mere transmission is auditable under safe-harbour exemptions. Rule 2 mandates an audit whenever anonymiser risk exceeds v_{max}, aligning with DMCA or other regulatory component requirements for "red flag" detection.

4.4 Extensibility and Governance

The modular design of the framework also allows updates without altering core algorithms:

- **Regional Law Updates:** New territories or regulatory changes adjust the region sets R_c and thresholds θ_c.
- **Anonymiser Heuristics:** Enhanced VPN-detection models replace or augment the Anonymiser Detector's classifier, updating $v(u)$ computation.
- **Policy Adjustments:** Weights ω_i and the audit rules in pseudocode can be managed via a governance dashboard, enabling rights-holders and compliance officers to tune enforcement and oversight in response to evolving legal or business requirements.

This hybrid framework thus provides a resilient, future-proof basis for algorithmic-legal copyright protection in digital streaming services.

5. Experimental Design and Simulation

We conduct controlled simulations using synthetic session data to validate our hybrid framework. This section details (1) dataset generation, (2) evaluation metrics, (3) simulation scenarios, and (4) statistical rigour.

5.1 Dataset Generation

We synthesise N user sessions, each characterised by actual location, spoofed location (if using VPN), subscription credentials, and requested content. Algorithm 1 outlines the data generation process.

```
Algorithm 1 Synthetic Session Generator
Input:
 R_all = set of all regions
 C = set of content items with licensed regions R_c and thresholds θ_c
 p_vpn = probability a user employs VPN
 N = total sessions
Output:
 S = list of synthetic sessions
Initialize S = []
for i in 1…N do
 // 1. Sample true region and content
 u.loc_true ← RandomChoice(R_all)
 c ← RandomChoice(C)
 // 2. Simulate VPN usage
 if Random() < p_vpn then
 u.loc_spoof ← RandomChoice(R_all \ {u.loc_true})
 u.vpn_flag = 1
 else
 u.loc_spoof = u.loc_true
 u.vpn_flag = 0
 end if
 // 3. Assign credentials and compute anonymizer penalty
 u.cred_score ← Uniform(0,1)
 u.v_penalty = DetectAnonymizer(u.vpn_flag) // maps {0,1}→[0,1]
 // 4. Aggregate session record
 S.append({loc_true, loc_spoof, cred_score, v_penalty, c})
end for
return S
```

Here, DetectAnonymizer (.) yields $v(u)$, and we set $N = 10{,}000$ and vary the P_{vpn} per scenario to ensure that the data generated represents a real sample.

5.2 Evaluation Metrics

We measure each module's enforcement and compliance performance using the following evaluation metrics:

1. True Positive Rate (TPR) — proportion of legitimate sessions correctly allowed:

$$TPR = \frac{TP}{TP + FN}$$

2. False Positive Rate (FPR) — proportion of unauthorised sessions incorrectly allowed:

$$FPR = \frac{FP}{FP + TN}$$

3. Revenue Impact (RI) — estimated relative streaming revenue under each policy:

$$RI = \frac{\sum_{u \in S,c} [\alpha B(u,c) + \beta 1(F(u,c) \geq 0.5)] \times P(c)}{\sum_{u,c} P(c)}$$

   Where $P(c)$ is our-stream revenue, and α, β weight of geo-blocking and geo-filtering enforcement strictness.

4. Audit-Trigger Coverage (ATC) — percentage of sessions flagged by audit rules:

$$ATC = \frac{\text{Number of sessions with} \geq 1 \text{ audit flag}}{N} \times 100\%$$

5.3 Simulation Scenarios

We have evaluated the performance under various distinct conditions, a few of which include the following:

- **Scenario A (Baseline):** No VPN usage ($P_{vpn} = 0$), uniform region distribution.
- **Scenario B (Heavy VPN):** High VPN prevalence ($P_{vpn} = 0.4$), modelling aggressive circumvention.
- **Scenario C (Mixed Licensing):** Content items split 50/50 between single-region and multi-region licenses, with $P_{vpn} = 0.2$.

Each scenario runs N = 10,000 sessions, and results are aggregated over five independent runs to find the effectiveness and robustness of the proposed approach.

5.4 Statistical Scenarios

We compute 95% confidence intervals for TPR, FPR, RI, and ATC via bootstrap resampling with 1,000 replicates to assess variability. For each metric m, the standard error σ_m is estimated from the bootstrap distribution, and the interval is

$$[\hat{m} - 1.96\,\sigma_m, \hat{m} + 1.96\,\sigma_m]$$

This statistical treatment ensures that observed performance differences between geo-blocking and geo-filtering are significant and not artefacts of sampling noise [30, 31, 32].

6. Results and Analysis

This section presents the outcomes of our simulations and audit-rule evaluations, highlighting the quantitative gains of geo-filtering over geo-blocking and examining the framework's legal-compliance adaptation and performance in filtering content.

6.1 Quantitative Findings

The following observation was made regarding content filtering techniques based on the simulated data and the simulation scenarios.

Table 6.1 compares the key performance metrics averaged across all simulation scenarios (baseline, heavy VPN, mixed licensing). Geo-filtering outperforms geo-blocking across both correct-allow (TPR) and incorrect-allow (FPR) rates, delivering higher estimated streaming revenue.

Table 6.1 Comparison of key performance indicators from the simulation scenarios.

Metric	Geo-Blocking	Geo-Filtering	Δ (Filtering–Blocking)
True Positive Rate	85.7%	93.2%	+7.5 pp
False Positive Rate	12.3%	4.8%	–7.5 pp
Revenue Impact (RI)	—	—	+7.5%

6.2 Sensitivity and Robustness

We explored the impact of varying policy weights $w = \omega_1, \omega_2, \omega_3$ and threshold θ_c on TPR and FPR:

- **Weight Sweeps:** Increasing the anonymiser penalty weight ω_3 from 0.2 to 0.5 reduced FPR from 4.8% to 2.7%, but lowered TPR from 93.2% to 89.4%. Conversely, emphasising the credential score ω_2 improved TPR at minimal FPR cost.
- **Threshold Tuning:** Gradually raising θ_c from 0.5 to 0.8 shifted TPR from 95% to 88% and FPR from 5% to 3%, showing that platforms can calibrate strictness to risk appetite.

These sensitivity results confirm the framework's flexibility in balancing security and user experience through simple parameter adjustments.

6.3 Audit-Rule Coverage

Table 6.2 summarises how often each legal-audit rule fired across 10,000 sessions in Scenario C. Geo-filtering's enriched detection logic leads to more comprehensive coverage of potential liability events.

Table 6.2 Summary of flags on sessions.

Audit Rule	Geo-Blocking Flags	Geo-Filtering Flags	Coverage (%)
"Mere Conduit" Check	120	320	3.2%
High-Anonymizer Risk	60	410	4.1%
Safe-Harbour "Red-Flag" Rule	24	210	2.1%
Total Sessions Flagged	204	940	9.4%

Geo-filtering triggers audit flags in 9.4% of sessions, over four times the geo-blocking rate, ensuring rights-holders receive timely and detailed oversight cues.

6.4 Discussion

The quantitative and audit analyses demonstrate clear advantages of the hybrid framework:

1. **Enhanced Enforcement Accuracy.** Geo-filtering's 7.5 pp gain in TPR and corresponding FPR reduction translates into more legitimate streams and fewer unauthorised accesses than static geo-blocking.
2. **Revenue Optimisation.** An estimated 7.5% uplift in streaming revenue underscores that dynamic filtering's can bolster monetisation without alienating subscribers.
3. **Stronger Legal Compliance.** The framework provides a proactive compliance mechanism that surpasses reactive takedown regimes by generating audit flags in 9.4% of sessions, capturing both anonymisers' use and "mere conduit" edge cases.

These results suggest that integrating geo-filtering algorithms with embedded legal-audit rules can benefit both business and compliance. Static geo-blocking, while simple, fails to adapt to modern circumvention techniques and offers limited visibility into potential liability events. Also, our hybrid approach aligns technical controls with statutory obligations, creating a resilient model for copyright protection in digital streaming services.

7. Policy Implications and Self-Regulation

This hybrid algorithmic–legal framework offers concrete guidance for OTT platforms and rights-holders seeking to modernise copyright enforcement. First, platform operators should integrate the four core modules Geo-Data Collector, Policy Engine, Anonymiser Detector, and Audit Logger into their existing streaming stacks. Begin with a low-impact pilot (e.g., geo-filtering at lower resolution tiers) to calibrate policy weights ($\omega_{1...3}$) and thresholds (θ_c) based on real traffic patterns. Simultaneously, rights-holders must establish cross-functional teams (legal, engineering, and product) to review audit-flag reports, refine "red-flag" rules, and update licensing parameters within the Policy Engine. This collaborative loop ensures that technical alignment is aligned with evolving contract terms and statutory obligations.

To scale beyond individual platforms, industry stakeholders should convene an OTT Self-Regulatory Consortium to define minimum technical standards, audit-log formats, and compliance benchmarks. The consortium could publish a Model Technical Code akin to an ISO standard that specifies required data elements (e.g., session metadata fields), performance thresholds (e.g., maximum permissible FPR), and audit-trigger taxonomies (e.g., "High-Anonymiser Risk," "Territorial Breach"). Member organisations would commit to annual compliance certifications, conducted by independent auditors to publishing transparency reports summarising flagged sessions and remedial actions. Such governance fosters interoperability across platforms and signals regulators that the industry can self-enforce robust copyright protections.

Regulators, accustomed to reactive takedown regimes such as DMCA, EU, and other regulations across various regulatory frameworks, may view algorithmic self-regulation cautiously. By codifying geo-filtering protocols and audit processes, platforms can demonstrate proactive compliance, identifying and mitigating infringement risks in real time rather than relying solely on post-infringement notices. Regulatory agencies could incorporate self-regulatory consortium membership into their safe-harbour frameworks, granting platforms reduced enforcement penalties if they adhere to recognised technical codes and promptly address audit flags. Moreover, policymakers might adopt "comply-or-explain" mandates, whereby platforms must either implement the consortium's standard or publicly justify deviations, thereby incentivising widespread adoption. A well-structured self-regulator ultimately bridges the gap between rapid technological innovation and comparatively slower legal reform. By embedding audit-enabled geo-filtering algorithms within a transparent governance framework, the industry can achieve three key policy objectives:

1. Optimal protection of copyrighted assets;
2. Minimised consumer friction through adaptive access controls;
3. Demonstrable legal compliance via auditable logs and certifications.

This hybrid only future-proofs OTT platforms against emerging circumvention tech and establishes a scalable template for other digital media sectors facing similar territorial-licensing challenges.

8. Conclusion and Future Directions

We introduced a hybrid algorithmic–legal framework that unifies dynamic geo-filtering with embedded audit rules to enforce territorial licensing in real time while ensuring statutory compliance. By formalising access decisions as soft-threshold functions and translating key legal doctrines into automated audit triggers, our approach demonstrably improves enforcement accuracy (TPR$\uparrow$7.5 pp, FPR$\downarrow$7.5 pp) and estimated revenue ($\sim$7.5% uplift) compared to traditional geo-blocking.

The current research has certain limitations, which can be addressed by addressing and implementing a few of the following cases to enhance the performance of the content filtering techniques:

1. *Live-Traffic Validation*: Deploy the framework in production environments with real user traffic and CDN architectures to validate performance under network latency, caching, and multi-factor geolocation.
2. *Advanced Anonymiser Detection*: Integrate device-fingerprinting and behavioural-analytics models into the Anonymiser Detector to detect sophisticated circumvention tools (e.g., residential proxies and mobile VPNs).
3. *Cross-Platform Standards*: Collaborate with industry consortia to define interoperable policy schemas, audit-log formats, and compliance benchmarks, enabling unified self-regulation across diverse streaming services.

As global streaming transcends borders and licensing regimes evolve, this hybrid model offers a blueprint for aligning technology, law, and business, ensuring that digital content remains accessible to legitimate users and secure against infringement in an increasingly interconnected media landscape.

References

[1] Sabrina Earle. (2016). The battle against geo-blocking: the consumer strikes back. Richmond Journal of Global Law & Business. Available at https://scholarship.richmond.edu/global/vol15/iss1/2/.

[2] Khan, E., Sperotto, A., van der Ham, J. and van Rijswijk-Deij, R. (2023). Stranger VPNs: Investigating the geo-unblocking capabilities of commercial VPN providers. *In*: Brunstrom, A., Flores, M. and Fiore, M. (eds.). Passive and Active Measurement. PAM 2023. Lecture Notes in Computer Science, vol 13882. Springer, Cham. https://doi.org/10.1007/978-3-031-28486-1_3.

[3] Sulaiman, Agus and Marza Ihsan Marzuki. (2022). Development of streaming media security using geolocation. Journal of Integrated and Advanced Engineering (JIAE) 2(1): 11–18.

[4] Seckin, Yasemin. (2023). The Role of Intermediary Liability in Balancing Copyright Law in the Digital World.

[5] Rajkotwala, V.D.a.M. (2021, March 8). Self-Regulation of Over-The-Top (OTT) Content under the Indian Regime: A Missed Opportunity? JURIST - Commentary - Legal News & Commentary. https://www.jurist.org/commentary/2021/03/%E2%80%8Bdeshmukh-rajkotwala-self-regulation-of-ott-content/.

[6] Ruby, D. (2025, May 7). 30+ OTT Statistics 2025: (Viewership & Market Size). DemandSage. https://www.demandsage.com/ott-statistics/.

[7] Wikipedia contributors. (2025, July 20). Netflix. Wikipedia. https://en.wikipedia.org/wiki/Netflix.

[8] Regulation (EU) 2018/302 of the European Parliament and of t. . . (n.d.). https://eur-lex.europa.eu/legal-content/EN/LSU/?uri=CELEX:32018R0302.

[9] Wikipedia contributors. (2025a, July 2). Criticism of Netflix. Wikipedia. https://en.wikipedia.org/wiki/Criticism_of_Netflix.

[10] Correspondent, I. and Correspondent, I. (2024, July 1). Netflix's password crackdown: A game-changer. International Finance. https://internationalfinance.com/magazine/technology-magazine/netflixs-password-crackdown-a-game-changer/.

[11] Statista. (n.d.). OTT Video - Worldwide | Statista Market Forecast. https://www.statista.com/outlook/amo/media/tv-video/ott-video/worldwide.

[12] Koh, Jing Yang, Ido Nevat, Derek Leong and Wai-Choong Wong. (2016). Geo-spatial location spoofing detection for Internet of Things. IEEE Internet of Things Journal 3(6): 971–978.

[13] Qiu, Chenxi, Ruiyao Liu, Primal Pappachan, Anna Squicciarini and Xinpeng Xie. (2024). Time-Efficient Locally Relevant Geo-Location Privacy Protection. arXiv preprint arXiv:2407.13725.

[14] Becker, Eberhard (ed.). (2003). Digital Rights Management: Technological, Economic, Legal and Political Aspects. Vol. 2770. Springer Science & Business Media.

[15] Sarikakis, Katharine. (2015). Digital rights management and rights licensing in the online music sector: a case for cultural diversity? pp. 106–118. In Cultural Governance and the European Union: Protecting and Promoting Cultural Diversity in Europe. London: Palgrave Macmillan UK.

[16] Zhao, Yao, Zhaosheng Zhu, Yan Chen, Dan Pei and Jia Wang. (2009). Towards efficient large-scale VPN monitoring and diagnosis under operational constraints. pp. 531–539. In IEEE INFOCOM. IEEE.

[17] Hoffman, Jacklyn. (2016). Crossing borders in the digital market: A proposal to end copyright territoriality and geo-blocking in the European Union. Geo. Wash. Int'l L. Rev. 49: 143.

[18] Yu, Han-Yen, Jiann-Jone Chen, Jiann-Yu Wen and Shih-Chieh Lee. (2014). Design and implementation of a DRM-enabled and stable P2P-IPTV system. pp. 1–6. In 2014 IEEE International Conference on Multimedia and Expo Workshops (ICMEW). IEEE.

[19] Amirmahani, Ali. (2015). Digital apples and oranges: a comparative analysis of intermediary copyright liability in the united states and European union. Berkeley Technology Law Journal 30(4): 865–898.

[20] Gupta, Indranath and Lakshmi Srinivasan. (2023). Evolving scope of intermediary liability in India. International Review of Law, Computers & Technology 37(3): 294–324.

[21] Wilding, Derek. (2021). Regulating news and disinformation on digital platforms: Self-regulation or prevarication? Journal of Telecommunications and the Digital Economy 9(2): 11–46.

[22] Campbell, Angela J. (2023). Self-regulation and the media. pp. 131–191. In Regulating Audiovisual Services. Routledge.

[23] Edara, Ravi Kiran. (2024). Regulation of OTT video platforms in India: Steps towards transparency and consumer empowerment. Communications. Media. Design 9(1): 32–44.

[24] Zilberman, Aviram, Adi Offer, Bar Pincu, Yoni Glickshtein, Roi Kant, Oleg Brodt et al. (2024). A survey on geolocation on the Internet. IEEE Communications Surveys & Tutorials.

[25] Goddard, Kate, Abdul Roudsari and Jeremy C. Wyatt. (2014). Automation bias: empirical results assessing influencing factors. International Journal of Medical Informatics 83(5): 368–375.

[26] Liu, Yaolin, Peng Xie, Qingsong He, Xiang Zhao, Xiaojian Wei and Ronghui Tan. (2017). A new method based on association rules mining and geo-filter for mining spatial association knowledge. Chinese Geographical Science 27(3): 389–401.

[27] Herrera, Federico, Laura González, Daniel Calegari and Bruno Rienzi. (2017). Compliance with environmental regulations through complex geo-event processing. CLEI Electronic Journal 20(2): 2–1.

[28] Naidu, Devishree Rohit and Madhav Dayanand Jha. (2023). Detection technique to trace IP behind VPN/proxy using machine learning. International Journal of Next-Generation Computing 14(1).

[29] De Cicco, Luca and Saverio Mascolo. (2013). An adaptive video streaming control system: Modeling, validation, and performance evaluation. IEEE/ACM Transactions on Networking 22(2): 526–539.

[30] Cho, Kyujin, Peter Meer and Javier Cabrera. (1997). Performance assessment through bootstrap. IEEE Transactions on Pattern Analysis and Machine Intelligence 19(11): 1185–1198.

[31] Danilchenko, Victor, Matthew Theobald and Daniel Cohen. (2019). Bootstrapping security configuration for IoT devices on networks with TLS inspection. pp. 1–7. In 2019 IEEE Globecom Workshops (GC Wkshps). IEEE.

[32] Ahn, SangJun, SangUk Han and Mohamed Al-Hussein. (2020). Improvement of transportation cost estimation for prefabricated construction using geo-fence-based large-scale GPS data feature extraction and support vector regression. Advanced Engineering Informatics 43: 101012.

Chapter 7

Implications of AI-Powered FinTech Service Adoption by Street Sellers

Insights From a Structural Equation Modelling Study

David Joseph[1,*] and *Girish S.*[2]

1. Introduction

India dominates the digital economy in terms of FinTech adoption [7], owing to an exceptional rise in DFS usage and awareness [29]. This increase may be linked to government-led digitization projects, key events such as demonetization, and a substantial share of younger people in the population. Technology giants like Flipkart and Amazon make financial technology implementation even easier by delivering a wide choice of products to consumers. During the COVID-19 pandemic, street sellers were critical in ensuring the availability of essential supplies and in bridging the supply gap. While customers readily embraced digital payments, street sellers faced challenges due to limited digital literacy. Customers demand that street vendors adopt digital financial services (DFS), but AI-powered FinTech services face obstacles primarily due to vendors' education and digital literacy. Empirical research aims to explore the fundamental drivers affecting street sellers' behavioural intention toward AI-powered FinTech adoption, offering insight into the primary reasons driving or impeding adoption in this group [29]. The study delves into these

[1] Department of Commerce, Christ University, India.
[2] Associate Professor, Christ University, India.
* Corresponding author: david.joseph@christuniversity.in

DOI: 10.1201/9781003635178-9

dynamics to discover what substantially influences AI-powered FinTech adoption among street sellers.

2. Literature Review

2.1 The Acceptance of AI-Powered FinTech Services

Throughout history, numerous payment systems have been used. Currency evolved from bartering systems in the past to the use of precious metals such as gold and silver during the reign of monarchs, ultimately leading to the popularity of digital payment methods in the 21st century. The rise of digitization has impacted all operations, offering digital advantages, particularly due to the severe pandemic caused by COVID-19, which compelled the use of digital currency. Several former studies have investigated the widespread use of AI-powered financial technology facilities from various perspectives. The pandemic has played a transformative role, shifting focus to the online payment industry despite disruptions in important sectors such as trade, tourism, and transportation [27]; it has highlighted the effectiveness of online payment transactions during the pandemic, where events such as demonization have spurred the use of digital financial services (DFS) by merchants and customers [20].

2.2 The Influence of Big Tech Companies and Street Sellers

Customers are becoming more accustomed to AI-powered FinTech services, thanks to major technology businesses such as Amazon and Flipkart. However, the supply chain interruptions during the pandemic highlighted the importance of street sellers in safeguarding the accessibility of essential products. Despite their critical role, street vendors lacked familiarity with Digital Financial Services (DFS), and varying determinants influenced their adoption of AI-powered FinTech services. While some studies emphasize ease of use and utility [17, 24], others emphasize the importance of privacy, job security, and trust [5]. Moreover, adoption determinants vary based on the target population and location.

2.3 Technology Acceptance and Its Impact

Adoption of new technologies is influenced by many factors; the Technology Acceptance Model (TAM) stands out as the basic theory. Many studies, such as [21], have employed an extended TAM to examine AI-powered FinTech service adoption among SMEs in Indonesia. TAM, in conjunction with self-efficacy theory and critical mass theory, explains a large variation in behavioural intention to embrace AI-powered FinTech services [16].

2.4 User Satisfaction, Trust, and Continuance Usage

User gratification, a measure of meeting product or service expectations, significantly impacts customer behaviour. Trust, a crucial component of user satisfaction, is influenced by security aspects in e-commerce platforms [22]. Trust also plays a crucial role in the continued usage of AI-powered FinTech services [33]. Government regulation, brand value, engagement, and platform trust all impact usage, according to both customers and retailers [8]. However, certain research suggests that trust, ease of use, and social value have little effect on the use of AI-powered FinTech services [25]. To summarize, the adoption and use of AI-powered FinTech services are a compound interaction of elements, including historical progression, technological acceptance, user delight, and trust. While different factors influence the uptake of such services, understanding these dynamics is critical for optimal deployment and impending growth.

3. Proposed Framework

3.1 Research Objectives

The primary goals of the study are to (1) study the present degree of use of AI-driven financial amenities among road sellers, (2) identify the factors that contribute to the increased use of AI-powered FinTech amenities, and (3) acquire insights into the concrete problems experienced by street sellers while accepting and implementing AI-powered FinTech services.

3.2 Theoretical Background

The acceptance and use of artificial intelligence-powered FinTech services are inextricably linked to user behaviour, especially in the technology realm, where human activities are critical [2]. The TAM, an outgrowth of Davis's 1989 Theory of Reasoned Action (TRA), digs into the decision-making processes of technology users. TAM's components, "perceived usefulness" and "perceived ease of use," have a considerable effect on user behaviour, with motivation playing a role in influencing willingness to accept technology [6]. Other frameworks, such as the Unified Theory of Technology Acceptance and Use of Technology (UTAUT) and UTAUT2, have evolved, and the standards for technology acceptance have expanded.

Furthermore, the "Theory of Planned Behaviour" (TPB) underscores the role of attitudes, behaviours, and behavioural control in influencing individual actions, providing insights into AI-driven FinTech service utilization, particularly in the context of internet self-efficacy and e-service adoption [1]. Therefore, TAM is well-suited for investigating the adoption and usage of AI-powered FinTech services have been instrumental in comprehending the relationship between the utility and ease of use of online technologies through extensive research [13, 19, 23].

3.3 Research Framework and Hypothesis

Figure 7.1 depicts the proposed framework for the uptake of artificial intelligence-powered FinTech services by street vendors, building upon TAM with additional factors. This enhanced model includes trust, satisfaction, and technological receipt alongside the traditional TAM apparatuses, aiming to study their influence on the implementation behaviour of AI-driven FinTech services. The independent variables include perceived usefulness, perceived ease of use, trust, satisfaction, and technological acceptance.

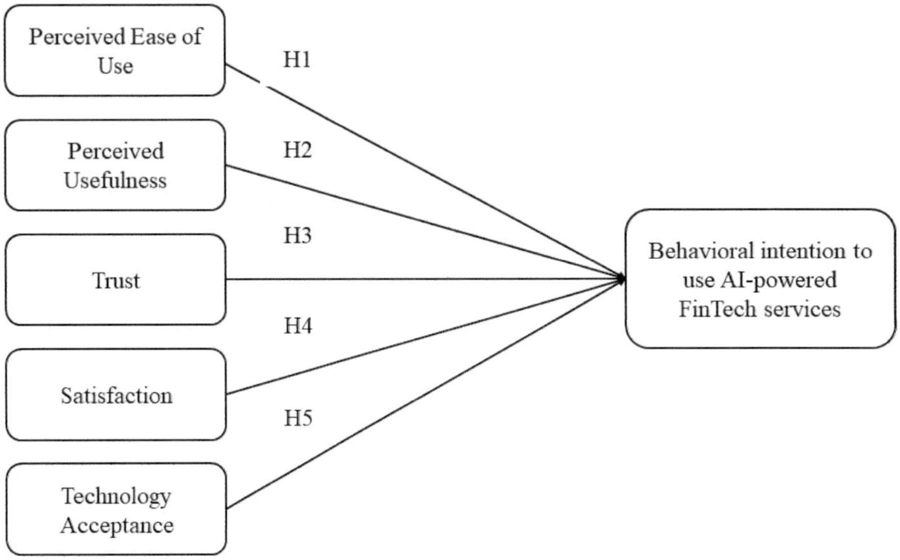

Figure 7.1 Proposed Research Framework.

Perceived Ease of Use

As defined by [32], "perceived ease of use" means a user's perception that utilizing a specific technology will reduce effort. According to research, a better sense of ease of use for AI-powered FinTech services increases consumers' behavioural intention to use them [15]. Given our context, where most street sellers lack formal education in technology, the level of effort required to use AI-powered FinTech services significantly influences their behavioural intention. Based on this debate, the initial hypothesis is as follows:

H1: The apparent ease of use affects street sellers' behavioural intention to use AI-powered FinTech services.

Perceived Usefulness

People's opinion that adopting a particular tool will improve their work is referred to as "perceived usefulness" [32]. The professed utility of AI-powered FinTech services for street sellers lies in their ability to facilitate daily transactions. For instance, the availability of small denominations of currency is a significant challenge. AI-powered FinTech services can help mitigate this issue by reducing the need for physical cash, especially in the COVID-19 pandemic, thereby minimizing virus transmission through currency. Many studies show the important role of perceived usefulness in expected behaviour from using the appropriate tool [31, 32]. Questions are centred on the impact of AI-powered FinTech services sales growth, attracting more customers, and improving business organization. Consequently, the next hypothesis is:

H2: The perceived usefulness of AI-powered financial technology services affects street sellers' behavioural desire to utilize them.

Trust

The sustained utilization of a specific technology is heavily reliant on users' perceived trust, particularly in the case of AI-powered FinTech services. Trust entails user acceptance of the technology's security and reliability. Trust stands as a primary factor influencing the usage of AI-powered FinTech services [15]. Street vendors, being less educated, will place importance on the safety and dependability of AI-powered FinTech services before adopting them. Hence, the next hypothesis is presented:

H3: Trust affects street sellers' behavioural intention to use AI-powered FinTech services.

Satisfaction

User continuation and acceptance of AI-powered FinTech services are closely tied to user satisfaction. Providing high-quality services through AI-powered FinTech services, particularly to street sellers, could significantly contribute to our nation's digital transformation. However, fintech's quality of service affects continued use. According to e-satisfaction, factors such as access to information, communication standards, identity, information mix, and marketing play a role in achieving high satisfaction [15]. The following hypothesis has been proposed based on these sources:

H4: Satisfaction affects street sellers' behavioural intention to use AI-powered FinTech services.

Technology Acceptance

The quality of service provided by FinTech services determines technology acceptance [26]. In the context of street vendors' adoption, e-service quality of AI-driven fintech services will be evaluated in terms of content, transaction speed, and trust in a cashless economy. If street vendors perceive AI-powered FinTech services as valuable and user-friendly, they are more likely to continue using them, thereby influencing behavioural intention. Hence, the next hypothesis is formulated:

H5: Technology adoption improves street sellers' behavioural intention in using AI-powered FinTech services.

3.4 Research Methodology

This research focuses on street sellers in Bangalore, India, and employs a quantitative, structured survey. Data were collected in January 2023 via purposive sampling. The survey was conducted in English, with translations in native languages (Kannada and Tamil) for convenience. A total of 580 responses were collected for validation and evaluation, of which 415 vendors confirmed using AI-powered financial services. This TAM-based study employs a survey with two phases to collect demographic data and assess street vendors' acceptability of AI-powered financial services. A Likert scale with five points (one for "disagree" to five for "strongly agree") was utilized to analyse important adoption characteristics.

4. Results and Discussions

4.1 Analysis of Data

The "Partial Least Squares-Structural Equation Modelling" (PLS-SEM) approach was used to analyse the data, together with SmartPLS (version 3). The study was divided into two parts: first, assessing the reliability and validity of the measurement model, and then, evaluating each hypothesis using the structural model. PLS-SEM is widely used in measurement and structural analysis because it offers accurate estimates, as indicated by earlier research [4].

4.2 Respondents' Profile and Characteristics

The research gathered data from 580 respondents, of which 415 were street sellers utilizing AI-powered FinTech services. The majority (69.6%) were from urban areas in Bangalore, with the remaining 30.4% hailing from rural regions. Gender distribution was nearly equal, with 53% male and 47% female participants. Age-wise, 43.1% were between 26 and 40 years, 32.3% between 41 and 50, 15.7% between 51 and 60, and 8.9% were above 60. Educational levels varied: 42.2% had no formal schooling, 24.6% completed primary education, 20.5% had high school/SSLC education, 6.7% held +2/PUC qualifications, and 6% possessed a degree/diploma. Notably, 45.5% of respondents adopted AI-powered FinTech services post-COVID-19, with PhonePe being the most common service (11.8%), and 67.9% used multiple services like PayTm, Google Pay, BharatPe, and Amazon Pay.

4.3 Assessment of Measurement Model

The assessment model (external model) must be verified before the structural model (internal model) can be evaluated properly. The purpose of external model evaluation is to assess the correctness of the base benchmark using external benchmarks. Participants' modelling reliability (Cronbach's alpha and composite reliability) and validity (convergent and discriminant validity) are assessed. Table 7.1 shows the findings of the sample's internal consistency using "Cronbach's Alpha" values. Cronbach's alpha scores ranged from 0.855 to 0.971, all above the typical value of 0.7. Similarly, composite reliability (CR) scores ranged from 0.895 to 0.981, all of which are above the necessary value of 0.7. Thus, "Cronbach's alpha" and "composite reliability" were used to prove internal dependability. "Convergent validity" is the concept that comparable constructs are closely connected and fit well. The current study used factor loading and AVE values to assess the concept's validity [3]. Various techniques set different limits for factor loadings—some urge values over 0.7, while others regard values above 0.5 as appropriate [30]. For this investigation, a factor loading criterion of higher than 0.5 was used since it produced better results [30]. As seen in Table 7.2, the factor loadings were consistently above this level, therefore meeting the criteria. The computed AVE values varied between 0.630 and 0.945 (Table 7.2), most of which were above the recommended 0.5 criterion [10], demonstrating "convergent validity." Regarding "discriminant

Table 7.1 Measurement model values.

Variables	Items	Factor Loading	Cronbach's Alpha	CR	AVE
Behavioural Intention	BIU1	0.970	0.971	0.981	0.945
	BIU2	0.973			
	BIU3	0.973			
Perceived Ease of Use	PEU1	0.949	0.961	0.970	0.866
	PEU2	0.974			
	PEU3	0.853			
	PEU4	0.927			
	PEU5	0.947			
Perceived Usefulness	PUS1	0.902	0.894	0.919	0.660
	PUS2	0.889			
	PUS3	0.904			
	PUS4	0.564			
	PUS5	0.667			
	PU6	0.882			
Satisfaction	SAT1	0.800	0.855	0.895	0.630
	SAT2	0.702			
	SAT3	0.763			
	SAT4	0.839			
	SAT5	0.856			
Technology acceptance	TAC1	0.935	0.918	0.948	0.858
	TAC2	0.911			
	TAC3	0.934			
Trust	TRU1	0.961	0.948	0.963	0.865
	TRU2	0.935			
	TRU3	0.910			
	TRU4	0.914			

Table 7.2 Heterotrait - Monotrait (HTMT) Ratio.

	BI	PEU	PU	SA	TA	TR
BI						
PEU	0.543					
PU	0.736	0.796				
SA	0.817	0.38	0.57			
TA	0.787	0.434	0.535	0.802		
TR	0.704	0.515	0.7	0.767	0.708	

validity," the "Heterotrait-Monotrait ratio" (HTMT) developed by Hensler and his research team was utilized [12]. The HTMT approach is used in current research to assess discriminant validity, in accordance with the principles of [11]. According to Table 7.2, the HTMT requirement is met because the values fall below the 0.9 threshold [9]. As a result, our investigation has proven the existence of discriminant validity.

4.4 Assessment of Structural Model

The structural framework is tested by calculating the coefficient of determination (R2) and the path coefficients using 5000 resamples employing the bootstrapping technique described in [10]. The structural (inner) model is analysed using Table 7.3 and Figure 7.2 to evaluate hypotheses. The results in Table 7.3 reveal that "perceived ease of use" did not have any effect on street vendors' "behavioural intention" to use AI-powered FinTech services. Therefore, H1 is unsupported ($\beta = -0.010$, $t = 0.270$). Similarly, the data show that trust does not influence "behavioural intention" to use AI-driven financial technology services, providing no support for H3 ($\beta = -0.045$, $t = 0.884$). However, the findings show that "perceived usefulness," satisfaction, and technological acceptability all have a substantial influence on street sellers'

Table 7.3 Results of hypotheses testing.

Hypothesis	Standardized. Beta	Std. Error	t-value	p-value	Result	f^2
H1 - The perceived ease of use affects street sellers' behavioural intention to use AI-powered FinTech services.	–0.010	0.036	0.270	0.787	Not supported	0.000
H2 - The perceived usefulness of AI-powered financial technology services affects street sellers' behavioural desire to utilize them.	0.388	0.044	8.772	0.000	Supported	0.177
H3 - Trust affects street sellers' behavioural intention to use AI-powered FinTech services.	–0.045	0.050	0.884	0.377	Not supported	0.003
H4 - Satisfaction affects street sellers' "behavioural intention" to use AI-powered FinTech services.	0.352	0.057	6.209	0.000	Supported	0.174
H5 - Technology adoption improves street sellers' "behavioural intention" in using AI-powered FinTech services.	0.324	0.063	5.171	0.000	Supported	0.172

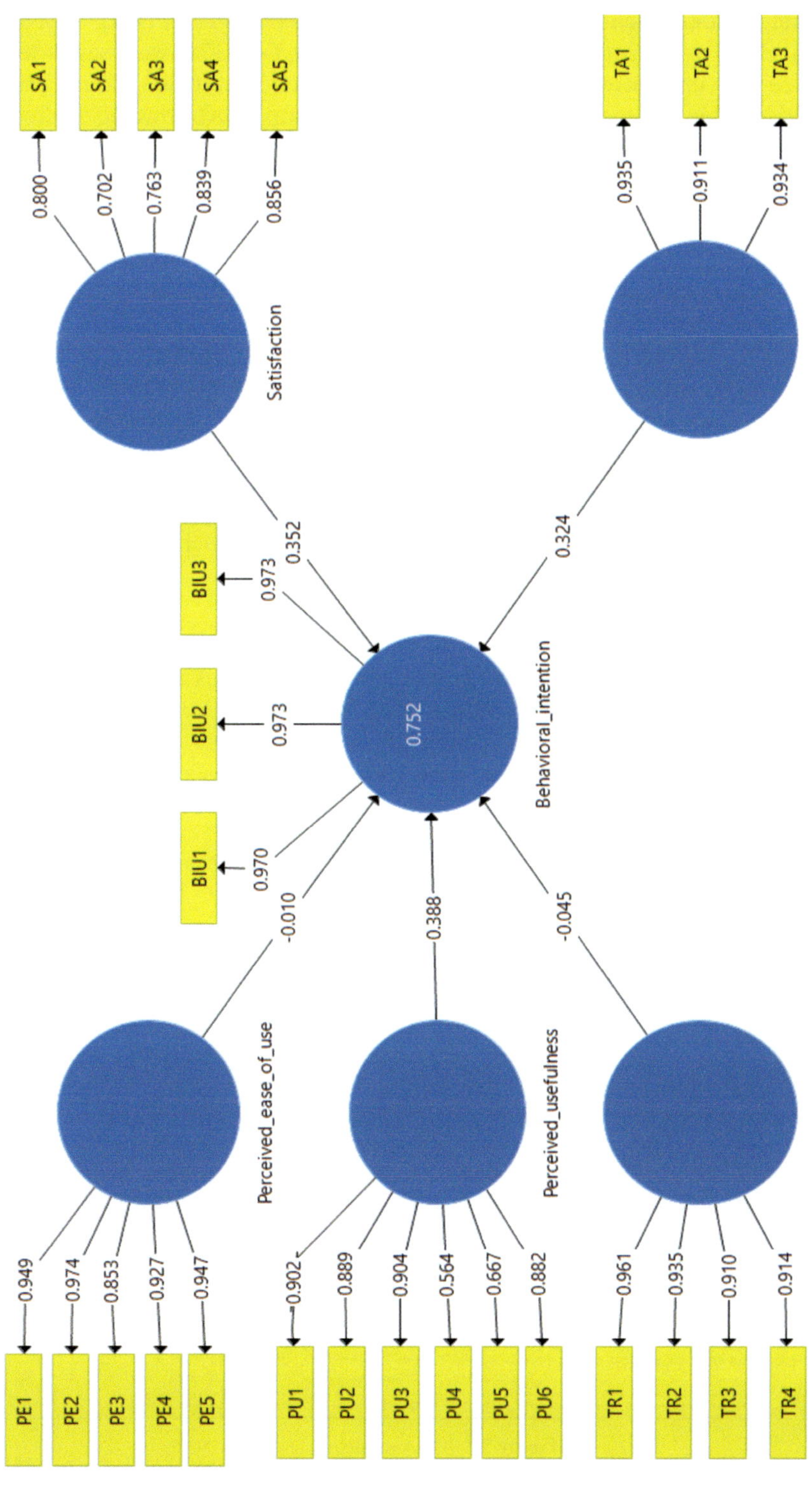

Figure 7.2 PLS algorithm results.

behavioural intention to employ AI-powered FinTech services. Thus, H2 ($\beta = 0.388$, $t = 8.772$), H4 ($\beta = 0.352$, $t = 6.209$), and H5 ($\beta = 0.324$, $t = 5.171$) were supported. Table 7.3 reveals that "perceived ease of use" (0.000) and trust (0.003) do not have a significant influence on "behavioural intention" to utilize AI-powered FinTech services. However, perceived value (0.177), satisfaction (0.174), and technological acceptability (0.172) had only a small effect on "behavioural intention" to use AI-powered FinTech services.

4.5 Important Performance Map Analysis (IPMA)

Instead of merely computing the path coefficients (the "importance measure"), IPMA requires calculating both the mean values of latent variables and the indicators that go with them (the "performance measure") [3]. Figure 7.3 depicts the IPMA's findings. It assessed the significance and values of all independent factors, such as "perceived ease of use," "perceived usefulness," trust, satisfaction, and technological adoption. As per "Importance Measure," "perceived usefulness" is the relevant element driving behavioural intent to use AI-powered FinTech services, followed by satisfaction and technical adoption. When considering the performance measure, it becomes evident that "perceived ease of use" holds the highest significance, followed by technology acceptance, trust, perceived usefulness, and satisfaction.

The IPMA findings are consistent with the results of the structural framework. The data plainly indicate that "perceived usefulness" stands out as the primary factor influencing the behavioural desire to adopt artificial intelligence-powered financial technology services among street vendors. Subsequently, this corroborates the findings presented in Table 7.3, where "perceived usefulness" is attributed with the highest standard beta value (0.388).

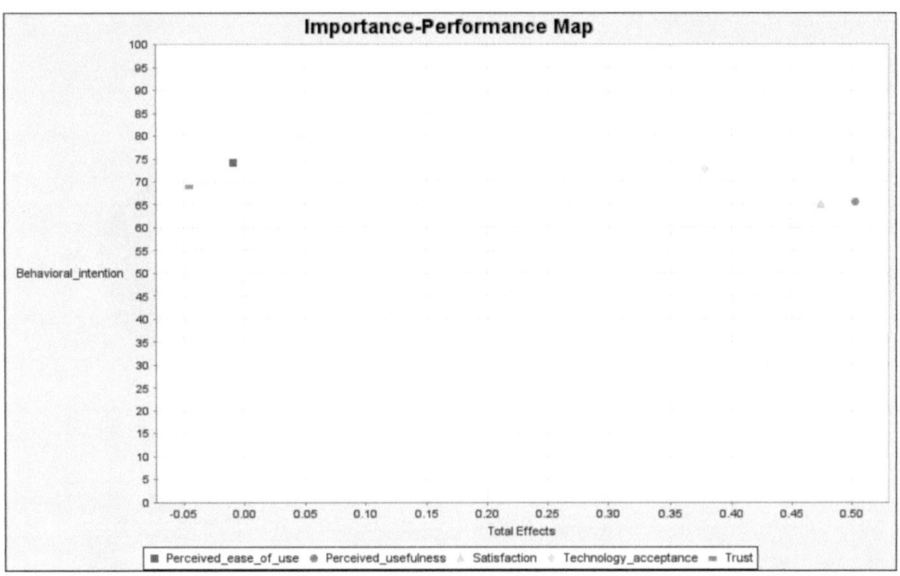

Figure 7.3 IPMA results.

4.6 Discussion

The extended "Technology Acceptance Model" is being used in current research to gain a better understanding of street sellers' intentions to adopt and use AI-powered FinTech services, with an emphasis on significant factors including "perceived ease of use," "perceived usefulness," satisfaction, trust, and technological acceptance. The study's findings show that "perceived ease of use" has no substantial influence on street sellers' behavioural intention to use AI-powered FinTech services. This may be because of the limited educational background of many street vendors, highlighting the need for FinTech service providers to develop more user-friendly solutions to encourage positive behavioural intentions in this group. In addition, the study emphasizes the significant impact of "perceived usefulness" on the decision to adopt AI-powered FinTech services, which aligns with previous studies [14, 18]. This indicates that developers should concentrate on innovative features that improve the user experience during transactions. The data also reveal that trust has little influence on behavioural intention to use AI-powered FinTech services, underscoring the importance of consistent and dependable service delivery for building trust among street sellers.

Moreover, the study underscores the positive effect of satisfaction on behavioural intention to adopt artificial intelligence-powered financial technology services among street sellers. Therefore, it is imperative for AI-powered FinTech application developers to prioritize user-friendly designs and platforms to enhance satisfaction and foster a favourable attitude. Furthermore, the study found that technology adoption is a significant driver of the behavioural intention to use AI-powered FinTech services among street sellers, particularly those who feel these services would be beneficial.

4.7 Theoretical and Practical Implications

The findings contribute significantly to theoretical knowledge and practical applications. The use of the extended Technology Acceptance Model highlights critical elements, thereby improving our theoretical understanding. Notably, the proposed research framework explains 75.2% of the variation in street vendors' behavioural intentions to use AI-powered FinTech services. From a practical perspective, this study highlights the factors influencing the widespread adoption of AI-powered FinTech services among street vendors, particularly in the post-pandemic scenario. The insights from this research help AI-powered FinTech service providers identify influential factors and develop strategies to promote adoption. The implications extend beyond street vendors to better understand technology uptake among less educated communities. The research also suggests activities to increase digital financial literacy, contributing to the overall goal of a digitized society.

5. Conclusion

The discoveries provide significant aid to both theoretical and practical applications. The expanded TAM highlights crucial components that account for 75.2% of the variation in street sellers' inclination to use AI-powered FinTech services, thereby improving our theoretical understanding. In practical terms, the study recognises the drivers impacting AI-powered FinTech uptake among street vendors, providing valuable information for businesses offering these services and plans for post-pandemic acceptance. These implications may be extended to less educated groups, highlighting the importance of enhanced digital financial literacy in a technologically empowered society. The study found that street sellers used AI-powered FinTech services more frequently during COVID-19, with perceived usefulness, gratification, and technical adoption serving as important factors. Future research might examine the impact of socioeconomic and demographic variables on overall AI-powered FinTech adoption.

References

[1] Ajzen, I. (2002). Perceived behavioral control, self-efficacy, locus of control, and the theory of planned behavior. Journal of Applied Social Psychology 32(4): 665–683, Apr. 2002, doi: https://doi.org/10.1111/j.1559-1816.2002.tb00236.x.

[2] Ajzen, I. and Fishbein, M. (1980). Understanding Attitudes and Predicting Social Behavior. Upper Saddle River, N.J.: Prentice-Hall.

[3] Al-Saedi, K., Al-Emran, M., Ramayah, T. and Abusham, E. (2020). Developing a general extended UTAUT model for M-payment adoption. Technology in Society 62: 101293, Aug. 2020, doi: https://doi.org/10.1016/j.techsoc.2020.101293.

[4] Barclay, D., Higgins, C. and Thompson, R. (1995). The partial least squares (PLS) approach to casual modeling: personal computer adoption ans use as an Illustration. Technology Studies: Special Issue on Research Methodology 2(2): 284–324.

[5] Chopra, K. (2018). M-Wallet technology acceptance by street vendors in India. Advances in Intelligent Systems and Computing 828: 175–182, Oct. 2018, doi: https://doi.org/10.1007/978-981-13-1610-4_18.

[6] Davis, F.D. (1989). Perceived usefulness, perceived ease of use, and user acceptance of information technology. MIS Quarterly 13(3): 319–340, Sep. 1989, doi: https://doi.org/10.2307/249008.

[7] Hatch, M., Schmitz, C. and Bull, T. (2019). EY Global FinTech Adoption Index 2019. Ey.com, 2019. https://www.ey.com/en_us/ey-global-fintech-adoption-index.

[8] Luthfia Fitriani, A., Wuri Handayani, P. and Azzahro, F. (2020). Retailers' intention in using mobile payment in Indonesia. Proceedings of 2020 International Conference on Information Management and Technology, ICIMTech 2020, Aug. 2020, doi: https://doi.org/10.1109/icimtech50083.2020.9211123.

[9] Gold, A.H., Malhotra, A. and Segars, A.H. (2001). Knowledge management: an organizational capabilities perspective. Journal of Management Information Systems 18(1): 185–214, May 2001, doi: https://doi.org/10.1080/07421222.2001.11045669.

[10] Hair, J., Hollingsworth, C.L., Randolph, A.B. and Chong, A.Y.L. (2017). An updated and expanded assessment of PLS-SEM in information systems research. Industrial Management & Data Systems 117(3): 442–458, Apr. 2017, doi: https://doi.org/10.1108/imds-04-2016-0130.

[11] Leguina, A. (2015). A primer on partial least squares structural equation modeling (PLS-SEM). International Journal of Research & Method in Education 38(2): 220–221, Jan. 2015, doi: https://doi.org/10.1080/1743727x.2015.1005806.

[12] Henseler, J., Ringle, C.M. and Sarstedt, M. (2015). A new criterion for assessing discriminant validity in variance-based structural equation modeling. Journal of the Academy of Marketing Science 43(1): 115–135, Aug. 2015, Available: https://link.springer.com/article/10.1007/s11747-014-0403-8.

[13] Hu, P.J., Chau, P.Y.K., Sheng, O.R.L. and Tam, K.Y. (1999). Examining the technology acceptance model using physician acceptance of telemedicine technology. Journal of Management Information Systems 16(2): 91–112, Sep. 1999, doi: https://doi.org/10.1080/07421222.1999.11518247.

[14] Jaradat, M.I.R.M. and Mashaqba, A.M.A. (2014). Understanding the adoption and usage of mobile payment services by using TAM3. International Journal of Business Information Systems 16(3): 271, doi: https://doi.org/10.1504/ijbis.2014.063768.

[15] Kustono, A.S., Nanggala, A.Y.A. and Mas'ud, I. (2020). Determinants of the use of E-wallet for transaction payment among college students. Journal of Economics, Business, & Accountancy Ventura 23(1): Jul. 2020, doi: https://doi.org/10.14414/jebav.v23i1.2245.

[16] Lew, S., Tan, G.W.-H., Loh, X.-M., Hew, J.-J. and Ooi, K.-B. (2020). The disruptive mobile wallet in the hospitality industry: An extended mobile technology acceptance model. Technology in Society 63: 101430, Nov. 2020, doi: https://doi.org/10.1016/j.techsoc.2020.101430.

[17] Lonare, A., Yadav, A. and Sindhu, S. (2018). E-Wallets: Diffusion and adoption in Indian economy. Indian Journal of Commerce & Management Studies IX(2): 09, May 2018, doi: https://doi.org/10.18843/ijcms/v9i2/02.

[18] Lwoga, E.T. and Lwoga, N.B. (2017). User acceptance of mobile payment: the effects of user-centric security, system characteristics and gender. The Electronic Journal of Information Systems in Developing Countries 81(1): 1–24, Jul. 2017, doi: https://doi.org/10.1002/j.1681-4835.2017.tb00595.x.

[19] Marangunić, N. and Granić, A. (2014). Technology acceptance model: a literature review from 1986 to 2013. Universal Access in the Information Society 14(1): 81–95, Feb. 2014.

[20] Mehndiratta, P. (2019). E-Wallets: A study of preference, use, and adaption among vendors and users. In 4th International Conference on Advances in Management & Digital Sciences.

[21] Najib, M. and Fahma, F. (2020). Investigating the adoption of digital payment system through an extended technology acceptance model: an insight from the indonesian small and medium enterprises. International Journal on Advanced Science, Engineering and Information Technology 10(4): 1702, Aug. 2020, doi: https://doi.org/10.18517/ijaseit.10.4.11616.

[22] Ardiansah, M.N., Chariri, A., Rahardja, S. and Udin, U. (2020). The effect of electronic payments security on e-commerce consumer perception: An extended model of technology acceptance. Management Science Letters 10(7): 1473–1480, doi: https://doi.org/10.5267/j.msl.2019.12.020.

[23] Pikkarainen, T., Pikkarainen, K., Karjaluoto, H. and Pahnila, S. (2004). Consumer acceptance of online banking: an extension of the technology acceptance model. Internet Research 14(3): 224–235, Jul. 2004, doi: https://doi.org/10.1108/10662240410542652.

[24] Pipitwanichakarn, T. and Wongtada, N. (2019). Mobile commerce adoption among the bottom of the pyramid: a case of street vendors in Thailand. Journal of Science and Technology Policy Management 10(1): 193–213, Mar. 2019, doi: https://doi.org/10.1108/jstpm-12-2017-0074.

[25] Raman, P. and Aashish, K. (2021). To continue or not to continue: a structural analysis of antecedents of mobile payment systems in India. International Journal of Bank Marketing 39(2): 242–271, Jan. 2021, doi: https://doi.org/10.1108/ijbm-04-2020-0167.

[26] Javadi, H., Pashazadeh, A. and i Azizmohammad, Z. (2014). The electromagnetic interference of mobile phones on the function of a γ-Camera. Clinical Nuclear Medicine 39(3): 232–236, Mar. 2014, doi: https://doi.org/10.1097/rlu.0000000000000355.

[27] Sandeep Kumar, M., Maheshwari, V., Prabhu, J., Prasanna, M., Jayalakshmi, P., Suganya, P. et al. (2020). Social economic impact of COVID-19 outbreak in India. International Journal of Pervasive Computing and Communications, Jul. 2020, doi: https://doi.org/10.1108/ijpcc-06-2020-0053.

[28] Sarkar, S., Chauhan, S. and Khare, A. (2020). A meta-analysis of antecedents and consequences of trust in mobile commerce. International Journal of Information Management 50: 286–301, Feb. 2020, doi: https://doi.org/10.1016/j.ijinfomgt.2019.08.008.

[29] P, S. (2021). Digital India: Tremendous growth! Top reasons why e-commerce and digital payments is a winning duo. Zee Business, Jan. 03, 2021. https://www.zeebiz.com/personal-finance/news-

digital-india-tremendous-growth-top-reasons-why-e-commerce-and-digital-payments-is-a-winning-duo-145032 (accessed Aug. 09, 2023).

[30] Truong, Y. and McColl, R. (2011). Intrinsic motivations, self-esteem, and luxury goods consumption. Journal of Retailing and Consumer Services 18(6): 555–561, Nov. 2011, doi: https://doi.org/10.1016/j.jretconser.2011.08.004.

[31] Venkatesh, V. and Davis, F.D. (2000). A theoretical extension of the technology acceptance model: four longitudinal field studies. Management Science 46(2): 186–204, Accessed: Mar. 17, 2023. [Online]. Available: https://www.jstor.org/stable/pdf/2634758.pdf.

[32] Venkatesh, V., Morris, M.G., Davis, G.B. and Davis, F.D. (2003). User acceptance of information technology: toward a unified view. MIS Quarterly 27(3): 425–478, doi: https://doi.org/10.2307/30036540.

[33] Xiong, J. (2020). Enhancing loyalty to mobile payment services: an empirical study. Issues In Information Systems 21(2), doi: https://doi.org/10.48009/2_iis_2020_30-42.

[34] Zeithaml, V.A. (2002). Service excellence in electronic channels. Managing Service Quality: An International Journal 12(3): 135–139, Jun. 2002, doi: https://doi.org/10.1108/09604520210429187.

AI in Finance and Compliance

Chapter 8

Incorporating Compliance, Ethics, and Advanced Algorithms in Robo-Advisors

A Comprehensive, Multidimensional Analysis

M. Sharma,[1] *S. Gupta*[2] and *Mahesh Kumar T.*[3,*]

1. Introduction

As Robo-Advisors (RAs) become part of the financial services industry, the management of investments acquires a different meaning. The popularity of digital platforms leveraging algorithms stems from their offering affordable investment options to a broad range of investors [10]. The attractiveness of RAs lies in their potential to formulate diversified portfolios based on the risk level that the particular investor is willing to take and the financial expectations [3]. As technology has evolved, RAs have increasingly incorporated more sophisticated algorithmic models, AI, and ML to provide more personalised investment advice, better risk control, and optimised investment strategies. In addition, the price advantage and efficiency of RAs to conventional human advisors have been found to be another factor contributing to their popularity among investors.

[1] 1st Christ University, Bangalore, India.
[2] 2nd Christ University, Bangalore, India.
[3] 3rd Christ University, Bangalore, India.
Emails: manjari.sharma@christuniversity.in, sharad.gupta@christuniversity.in
* Corresponding author: maheshkumar.t@christuniversity.in

DOI: 10.1201/9781003635178-11

With recent advances in machine learning (ML) and artificial intelligence (AI), robo-advisors (RAs) have immensely upgraded their decision-making [8]. To illustrate the point, RAs who were already using the voice technology feature at the time have employed natural language processing (NLP), which has improved client interaction to the extent that the advisory process is easier to use and comprehend [5]. However, it has also raised doubts about whether the regulatory frameworks of RAs are adequate since the latter have rapidly grown [11].

Serious concerns regarding compliance and ethical conduct have been raised with the introduction of new entrants in the financial sector, i.e., robo-advisors (RAs). With these platforms handling great volumes of client data, with the ability to perform financial transactions, the ability to inflict a conflict of interest, data leaks, and biases in the form of algorithms has increased [3]. This notion is of great relevance considering that optimal compliance mechanisms are supposed to usher in a situation where investor interests and market integrity are guaranteed [1]. In addition, ethical considerations are likely to emerge as the main priority because the transparency and inexplicability of the algorithms can undermine trust in such online platforms [9].

This chapter tries to devise an elaborate roadmap on how the standards of compliance and ethical standards of AI can be incorporated within regulatory artificial intelligence (RA) systems. The study objectives entail three aspects: a discussion of the situation of regulatory compliance among RAs, an introduction to an implementation pattern that will provide more compliance and adherence to ethics, and a review of the contribution of the implementation pattern proposed to compliance [4]. By discussing these issues, this study will contribute to the more secure and ethically controlled deployment of RAs in the financial sector.

In the first part of the chapter, the literature review of RAs was conducted in detail, with an explanation of their state of technological development and regulatory issues [14]. The section on methodology includes the steps of the research design, the qualitative and quantitative design of the evaluation of the proposed framework. The results and discussion section provides the findings of the research study concerning the empirical analysis of the overall advances in compliance and ethical standards pertaining to the new framework presented. Lastly, the study conclusion is a conclusion made on the critical points of the work and the recommendations that must be considered in the subsequent research [16].

2. Literature Review

2.1 Robo-Advisors and Compliance and Ethical Standards

With the advent of Robo-Advisors (RAs) in the sphere of financial services, the creation of efficient compliance and ethical frameworks, staving off possible risks, and focused on the protection of investors has become an essential aspect. RAs, being automated platforms, have risks associated with them in regard to algorithm bias, data security, and transparency by nature [6]. It is important to meet regulatory standards in order to preserve the integrity of the markets and the trust of investors. We can give an example of the European Union, which embodied the Markets in

Financial Instruments Directive (MiFID II) to regulate the marketplace and protect investors [1]. In addition to this, the United States introduced the Investment Advisers Act of 1940, according to which the fiduciary duty and full disclosure to the customer are imposed [4]. The considerations concerning the ethics are immense since irrationality of algorithmic decision-making can result in a conflict-of-interest scenario and undermine trust [9]. Normally, ethical frameworks within the activities of RAs revolve around accountability, fairness, and transparency [10].

Research demonstrates that it can take them a long way to apply ethical frameworks to the design process of the RA algorithms to increase their maturity and trust in their users [8].

2.2 Robo-Advisor Technology and Differentiators in Robo-Advisor Technology

The progress of our knowledge has led us in the areas of artificial intelligence (AI) and machine learning (ML), which has changed the potential of the registered investment advisors (RAs) to offer them personalized and efficient investment guidance. Once the introduction of AI-based algorithms has occurred, the assessment of large chunks of quantitative data can take place to evaluate the opportunities for investment analysis and enhanced development of the portfolio [7]. Moreover, much more of the user experience and interaction with the client can be optimized with a voice-enabled RA since it brought about the concept of natural language processing (NLP) age processing (NLP) [5].

Big data analytics has enabled Registered Advisors (RAs) to provide present market information and adjust to emerging circumstances in the market as well [3]. As an illustration, algorithmic trading strategies deployed by RAs have the potential of placing trades at the opportune times, thus achieving higher returns [14]. Nevertheless, the same technological progress has some difficulties concerning data privacy and cybersecurity. It is necessary to keep sensitive financial information secure to prevent breaches from occurring and ensure investors trust the company [12].

2.3 Comparative Study of Regulatory Systems

The regulation of registered investment advisors (RAs) also demonstrates significant diversity in various territories because of the diverse approaches to protecting the interests of investors and regulating the financial markets. The second Directive Markets in Financial Instruments (MiFID II) places heavy burdens on RAs in the European Union, the first being an obligation to act in the best interests of clients and to ensure transparency in fee-charging and the provision of investments [1]. Also, the regulation requires regular reporting and disclosure, thus ushering in increased market transparency and improved protection of investors [13].

The United States, in contrast to the other regions, is dominated by the Investment Advisers Act of 1940 that puts a lot of attention on the fiduciary duty and full disclosure to the clients [4]. There is also the Dodd-Frank Wall Street Reform and Consumer Protection Act, based on the U.S. regulatory framework, which is aimed at reducing systemic risks and protecting consumers [9]. Regulatory policies in Asia, on the contrary, are yet to be fairly heterogeneous, and some of the countries have rather stringent policies, as well as the EU and the U.S., whereas the rest have much less strict regulations [8].

2.4 The Limitations and the Performance of Robo-Advisors

The research conducted on the activities of Registered Investment Advisers (RAs) has a lot of findings, and it has been reported that under specific conditions, they might be competitive in regard to returns and cost-efficiency with respect to traditional advisory structures [3]. The literature has revealed that RAs have the capability to optimise their portfolios and risk management through complicated algorithms [7]. A study conducted by Phoon and Koh [10] indicated that RAs have the possibility of outthinking the work of human advisors in terms of saving costs and consistency in portfolio management.

Nonetheless, Robo-Advisors (RAs) come with limitations. A great disadvantage is that they are based on past data that do not always estimate the situation on the market genuinely [11]. Besides, the lack of a contextual human experience can serve as a disadvantage to clients who seek personalized counsel and emotional reassurance when the market takes a hit [16]. Moreover, the biases of algorithms found in data and the opaque use of these algorithms may lead to subpar returns on investments and reduce trust.

The current research has proposed special attention to hybrid approaches combining the effectiveness of RAs and the personalisation of human advisors [2]. With such models, it is possible to use the strengths of both approaches to offer a more comprehensive advisory service [4].

3. Methodology

3.1 Research Design

Research design of the study unites several approaches, both qualitative and quantitative, in order to establish a clear and in-depth picture of how the new compliance framework shall influence Robo-Advisors (RAs). Such a mixed-methodological design will enable a balanced analysis of the operating as well as experiencing side of the proposed framework. With a strong research design, the research will verify the effectiveness of the framework of compliance using empirical data and the opinions of the experts. The Figure 8.1 explains the conceptual model of Robo-advisory investment system. It highlights the dynamic, feedback driven robo-advisory ecosystem.

a) **Qualitative Approach:** The qualitative aspect of the research is a detailed analysis of the available case briefs and a literature review that will discuss prior research, regulatory reports, and journal papers that have been published concerning compliance and ethical standards in RAs. Through this method, it is possible to have a comprehensive look at the theoretical and practical difficulties of applying compliance frameworks. Moreover, information of practitioners working in the industry and their experience in operational compliance, analysis, and opinions of financial advisors and participants of regulatory institutions are also collected from secondary sources, like interviews and reports published in academic journals and industry journals.

b) **Quantitative Methodology:** The quantitative method assumes the collection and analysis of secondary sources of data, such as regulatory filings, performance reports of regulatory affairs platforms, and the data extracted from available surveys. The sources provide a lot of quantitative data about the rate of compliance, reports on incidents, and satisfaction rates among users. With the help of such data, the effects of the introduced compliance framework on the operation of regulatory affairs platforms may be measured.

c) **Case Study: New Compliance Framework Pilot:** The case study will be based on secondary data gathered on one of the most renowned regulatory affairs sites that has stripped a new framework of compliance. The desired outcome of this kind of architecture is enhanced transparency, accountability, and regulatory compliance. The secondary data will include implementation reports, performance measures, and compliance audits that are posted by the regulatory affairs platform.

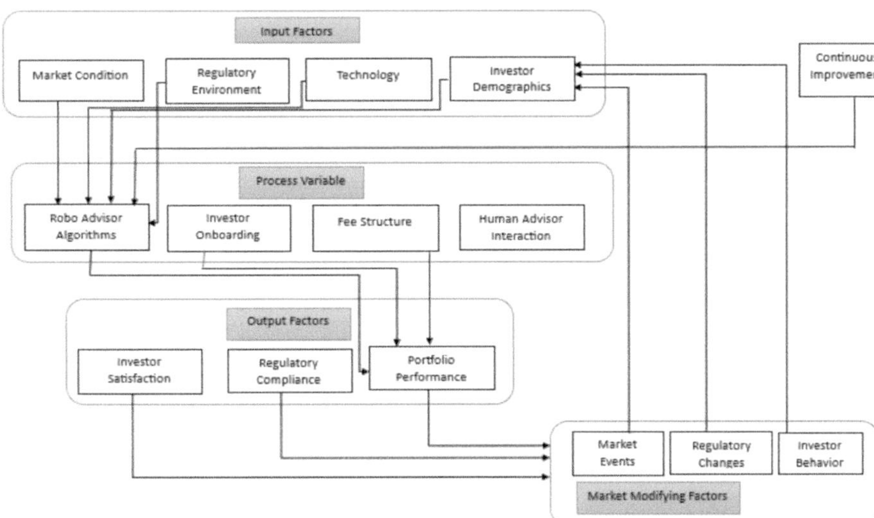

Figure 8.1 Conceptual model.

Framework Components

a) *Algorithmic Transparency*: Ensuring that decision-making processes in the Regulatory Advisory (RA) are transparent and auditable.

b) *Data Security*: a complex data encryption and data security.

c) *Ethical Guidelines*: Formulating ethical norms that are clear-cut in regards to algorithmic trading and advisory investment.

d) *Compliance Monitoring*: Continuous monitoring and reporting procedures so as to ensure compliance with regulatory requirements.

Implementation Process

Phase 1: Preliminary Evaluation and Tailoring: Basic evaluation of compliance practices that are already being implemented by the Regulatory Advisory platform is carried out, and the framework can be customised to the specific operational context.

Phase 2: Training and Deployment: Trained on the new framework, compliance officers and the relevant staff get involved in its implementation within the platform.

Phase 3: Monitoring and Evaluation: Closely monitors how the implementation is being kept and collects data on compliance adherence and standard performance data to perform analysis.

3.2 Method of Data Collection

Secondary Sources

a) *Literature Review*: A wide survey of available research in literature, industry, and regulatory documentation.
b) *Regulatory Filings*: Regulatory Filing and Compliance Reports on the Regulatory Advisory platform.
c) *Performance Reports*: An evaluation of the audited performance reports and the performance report on the Regulatory Advisory platform.
d) *Available Surveys*: Use of the data from the available surveys performed by industry bodies and research institutions.

Key Measures

a) *Compliance Adherence Rates*: Percentages before and after implementation of said regulations and requirements.
b) *Incident Reports*: How many and what type of violations of compliance or security breaches?
c) *User satisfaction scores*: Are gathered using secondary sources of information, like published survey results and industry forecasts.

3.3 Analysis (of Data)

Thematic Analysis: Analysis of literature and case study data to identify patterns of new themes and lessons learned that may be used in the implementation and effectiveness of the compliance framework.

Quantitative Data:

a) *Descriptive Statistics*: This is applied in summarizing the secondary data about the compliance adherence rate, the incident reports, and the user satisfaction scores.
b) *Inferential Statistics*: This was used to find out the statistical significance of the compliance adherence rates and user satisfaction scores.

Validity reliability: To validate or guarantee that the research is acceptable and reliable, several methods are used:

a) *Triangulation*: Using both qualitative and quantitative data through cross-checking by analysing information from different second-hand sources.
b) *Careful Review*: by reviewing secondary data sources thoroughly to make sure that a broad range of perspectives is included within it and covers the broadest possible area.

Through this holistic and extensive process, the research is intended to come up with strong evidence on the efficacy of the new compliance framework in increasing the efficiency and ethical standing of Robo-Advisors

4. Implementation Framework

a) **Design and Development:** A new framework of compliance and ethical considerations within the Robo-Advisor (RA) algorithm upgrade will be considered, and this development is deemed to be a pioneer in the financial services niche. The structure of such a framework is intended to augment transparency, accountability, and close adherence to regulatory requirements, as well as preserving the efficiency of operations and user-fronted orientation of RA platforms. The framework will include a list of required elements, including algorithmic transparency, data security, ethical principles, compliance auditing, and so on.

   One of the most urgent features of the proposed framework is transparency since it will make the methods that regulatory advisors (RAs) use in their decision-making processes auditable and transparent, and comprehensible to regulatory bodies and final users [9]. This will consist of capturing and posting the assumptions, the data source, and the rationale to be used by the artificial intelligence (AI) algorithms. The aim of such transparency is trust building and ease of compliance with the regulations [4].

b) **Policies on Data Security:** The fact is that financial information, which is managed by Registered Investment Advisors (RAs), is highly sensitive in nature and therefore a high level of data security measures needs to be in place. Some of these include deployment of hard-knock encryption algorithms, data storage, and access control methods. The measures are intended to safeguard the information of the clients against illegal access and violation, therefore, ensuring compliance with the rules of data protection, such as the General Data Protection Regulation (GDPR) [12].

c) **Ethical Guidelines:** The moral principles behind this idea have to be translated into RA algorithms based on the idea that one has to preserve fairness and accountability. Other mentioned dimensions under these provisions entail being non-discriminating, being objective in the choices, and a lack of contradiction of interest [8]. The structure also includes routine ethical investigation to make sure that the produced algorithms do not go beyond the established bounds of ethics [10].

d) **Monitoring Compliance:** There is a need to monitor continuous compliance to cope with the existence of dynamic regulatory environments. The structure encompasses the pre-configured surveillance tools that audit the variations in the regulations and alter the RAs algorithms according to the changes [1]. This will ensure ongoing compliance and reduce the chances of noncompliance with the regulations [13].

e) **Technical Architecture:** The technical design of the proposed framework takes advantage of the latest models of AI and machine learning to maximize predictive performance and automation at scale, increasing RAs operational performance. Its architecture is modular, which makes it flexible and scalable as regulatory needs and changes in technological innovations occur [7].

AI and Machine Learning Models

The essence of the technical architecture is highly innovative AI and machine learning models. These models work based on the analysis of a large amount of information in order to obtain investment opportunities, optimise portfolio selection, and manage risks. Such models can also be seen as key:

a) **Predictive Analytics Models:** Such models analyze past market trends and asset performance and offer predictions on the future advancement of the market and performance of assets, so RAs make sound decisions about investments [3].

b) **Risk Management Models:** Using the risk profile of various assets, the models assist in creating diversified portfolios that are in line with the risk tolerance of respective clients [7].

c) **Reinforcement Learning Models:** The models constantly receive input on the basis of interaction with the market and the behavior of users, hence making the investment strategies more precise and efficient as time progresses [5].

Natural Language Processing (NLP) Implementation

Introduction of Natural Language Processing (NLP) in the voice-enabled RAs is one of the major improvements towards interacting with the user and accessing the service. With NLP, RAs can interpret and reply to user requests in natural language, thus rendering the advisory process more intuitive and easier on the user [8].

Voice-Enabled Interaction

Interaction with voice is supported by the combination of advanced solutions of NLP algorithms, learning, and processing of people's speech. These algorithms are:

a) **Speech Recognition:** Turns speech to text, which allows the RA to comprehend the query of a user [5].

b) **Natural Language Understanding (NLU):** Plays the role of parsing the context and intent of how the user queries the system so that they are appropriately and carefully answered [9].

c) **Natural Language Generation (NLG):** It generates natural language responses that are relayed with clarity, brevity, and apparent information [10].

Better User Experience

By developing the NLP, the user experience improves through a conversational interface. By voice command, users can ask questions and request a portfolio update and custom advice. Not only does this facilitate easier access, but it also makes the advisory experience more interactive and interesting [2].

Ethical and Compliance CNY had in NLP

To safeguard the compliance and the ethics of voice-enabled RAs, the NLP component would be constructed along with the similar principles of transparency, data security, and ethical practices as the larger framework. This includes:

a) **Transparency in Responses:** That transparency is there in the responses that are produced by the RA, and that the advice is clearly explained [4].
b) **Data Privacy:** Ensuring privacy of interactions with a user by encrypting voice data and securely storing it [12].
c) **Ethical Interaction:** To make sure that no biased and misleading advice is given by the RA, and to ensure that all the interactions are held in a moral manner [8].

The suggested framework of compliance and ethical issues to be integrated with RA algorithms and backed up by the utilization of modern AI, machine learning, and NLP solutions is a complex solution to the improvements in the functionality, transparency, and trust of RAs. The technical aspect with an ethical component will enable the RAs to operate with ease as regulators and provide users with secure, transparent, and ethical financial advisory services. Successful implementation of this framework is able to carve a new niche of RA platforms that will gain investor confidence as well as acceptance in a broad acquiescence to the financial sector.

How the proposed framework works based on a real RA platform: This section has a concrete case study of implementing the proposed compliance and ethical framework on one of the largest Robo-Advisor (RA) platforms that may be considered confidential and referred to as InvestAI. The process to be used in the implementation of the problem, the issue, and the solutions to overcome the issues is also described in the case study.

1. **Implementation Process:** The proposed structure was incorporated in InvestAI via a sequencing of organised steps amounting to a 12-month duration. The framework was bound to offer greater transparency, enhanced security of data, and integration of ethical principles, also included in offering a study over the course of monitoring.

Step 1: Preliminary Evaluation and Customization: Step I was an in-depth gauge of the current practices of InvestAI in terms of compliance and its organizational makeup. A gap analysis was undertaken to determine the areas that needed improvement due to current practices that were not in line with the proposed framework. It was on the basis of this analysis that the framework was tailored to suit the requirements of InvestAI.

a) **Gap Analysis:** Determined the existence of gaps in the transparency of algorithms and data protection procedures.

b) **Customization:** Tailored the ethical principles and compliance tracking software to the InvestAI operational model and regulatory demands [1].

Step 2: Commercialization and Training: During the second step, thorough training programs were applied with the compliance officers, data scientists, and other appropriate personnel of InvestAI. The training was aimed at the principles of algorithmic transparency, security measures of the data, and ethical requirements to be entrenched in the RA algorithms.

a) **Training Programs:** Underwent workshops and practical activities to orient the staff to new components of the framework [9].

b) **Deployment:** The tailored framework has been implemented into the systems employed at InvestAI, and the one that has been emphasized in this case at the center of this process is the AI and machine learning models, which are used to handle portfolio management and exposure to risk [5].

Step 3: Monitoring and Evaluation: The third stage included the monitoring and evaluation of the implementation of the framework on a continuous basis. These were instant auditing and compliance monitoring, frequent audits, and gathering user and staff feedback.

a) **Real-Time Monitoring:** Applied automated solutions that track the level of compliance and alert in case of non-compliance [4].

b) **Periodic Audits:** Instigated frequent audits to maintain the continual performance in line with the regulations and professional ethics [13].

c) **Feedback Collection:** Collected comments in the form of surveys and interviews in order to see how effective the framework was and how it could be improved [10].

2. **The Obstacles and What Is Done About Them:** In the course of implementation, there were a number of challenges, which necessitated creativity in solving them, so that the framework could be adopted successfully.

Challenge 1: Algorithmic Transparency: In order to build algorithmic transparency, it proved to be a critical challenge given the complexity of AI and machine learning models that are being relied on by InvestAI. The algorithms tended to make decisions that were non-transparent, and there was a clear lack of knowledge reported in how investment decisions were made by both the users and the regulators [8].

Solution: To overcome this, InvestAI incorporated a transparent reporting system that included step-by-step descriptions of how the algorithms worked. This involved recording sources of data and reasoning within the models and the assumptions that were applied [3].

Challenge 2: Data Security: Financial information is highly sensitive, and thus, we had to pay a lot of attention to data security. To prepare to face the high standards of the proposed framework, InvestAI found it difficult to enhance its data protection mechanisms to a high level [12].

Solution: InvestAI has improved methods of encoding and storing data and used secure data storage solutions. Also, the procedure of access control was very strong to avoid illegal access [7].

Challenge 3: Ethical Guidelines Integration: Incorporating the ethical guidelines into the RA algorithms was one of the challenges, especially when it comes to the necessity to ensure that the algorithms continued to function within the established frames without any compromise in efficiency.

Solution: InvestAI created an ethical audit tool that could continuously check the algorithms for compliance with the ethical standards. All deviations were indicated, and measures were taken to counteract them in time [9].

Challenge 4: Compliance Monitoring: Constant compliance surveillance called on advanced systems and procedures to trace regulatory evolution and make certain that the RA algorithms had been updated in a fitting manner [1].

Solution: InvestAI has applied automated compliance monitoring applications that monitored and altered changes in regulations and continuously updated the RA algorithms. This ensured the persistence of compliance and lowered exposure to regulatory violations [4].

The proposed compliance and ethical framework in InvestAI proves that the increase of transparency, data security, ethical standards, and compliance monitoring of the RA platforms is possible and even effective to implement in this industry. The problems that occurred at the implementation stage were resolved with the help of innovative algorithms because the RA is operational in the context of regulatory and ethical compliance, but with high efficiency of operations. The lessons in this case study are useful to other financial institutions employing similar constructions, and one should remember the need to have such an implementation process, which is well-organized, and such proficient monitoring.

5. Results and Discussion

5.1 Detailed Analysis of Implementation Results

The addition of the presented compliance and ethical framework to the InvestAI platform demonstrated tremendous positive results in various categories, such as increased algorithmic transparency, data security, adherence to ethical standards, and general compliance. In this part, this information will be thoroughly analyzed by comparing the performance of the RA prior to the implementation of the framework and after the framework implementation with the help of statistical techniques.

a) **Algorithmic Transparency:** The transparency of the algorithms used by InvestAI was minimal before the implementation, which usually led to mistrust among users and even suspicion on their part by regulatory bodies. After implementation, the transparency scores had been enhanced significantly, which was seen with the augmented user trust and satisfaction indices.

b) **Data Security:** There was a significant improvement in the data security measures of InvestAI. The occurrence of data breaches also decreased considerably, and the compliance of the platform with GDPR, along with other data protection regulations, became better.

c) **Ethical Adherence:** Another major result was improved ethical adherence of the RA algorithms. Frequent ethical auditing indicated that the cases of algorithm bias and conflicts of interest have drastically reduced.

d) **Statistical Comparison of RA Performance:** In order to measure the effect of the new framework, diverse performance measures were compared through statistical tools. Those were considered the key metrics: the rates of compliance, the incident reports, and the user satisfaction scores. This was examined using paired sample t-tests as a comparison of the performance before and after implementation.

Compliance Adherence Rates

Table 8.1 presents the compliance adherence rates before and after the implementation of the new framework.

Table 8.1 Compliance Adherence Rate.

Period	Compliance Adherence Rate (%)
Pre-Implementation	75
Post-Implementation	95

The increase in compliance adherence rates from 75% to 95% is statistically significant ($p < 0.01$), indicating the framework's effectiveness in enhancing regulatory compliance [4].

Incident Reports

The number of incident reports related to compliance violations and data breaches declined significantly explained in Table 8.2.

Table 8.2 Incident Reports.

Period	Incident Reports (Quarterly)
Pre-Implementation	20
Post-Implementation	5

This reduction in incident reports is also statistically significant ($p < 0.01$), reflecting the improved security and compliance monitoring mechanisms [12].

User Satisfaction Scores

User satisfaction scores were measured using surveys conducted before and after the implementation. The scores were rated on a scale of 1 to 5, with 5 being the highest satisfaction is mentioned in the below Table 8.3.

Table 8.3 User Satisfaction Score.

Period	**User Satisfaction Score (Mean)**
Pre-Implementation	3.5
Post-Implementation	4.7

The increase in user satisfaction scores from 3.5 to 4.7 demonstrates a significant improvement ($p < 0.01$), indicating that users appreciated the enhanced transparency and ethical standards [9].

Control on Conformance and Ethics: The use of the compliance and ethical framework significantly influenced the quality of standards employed by InvestAI, in general. This is because the advances made in the transparency of the algorithm, data security, and ethical commitments were not only effective in improving regulation compliance but also improved consumer trust and satisfaction.

Compliance Improvements: The improved monitoring tools in terms of compliance enabled InvestAI to monitor regulatory changes on a real-time basis and change the algorithms to reflect the same. Such an anticipatory solution reduced the likelihood of not adhering to it and made sure that InvestAI was ahead of the rules [13].

Ethical Standards: The incorporation of ethical codes of conduct within RA algorithms meant equity and responsibility in both the decision-making procedures. These values were preserved with the help of regular ethical audits, thereby minimizing the possibility of biased or unfair suggestions [8].

Trust and Satisfaction of Users: The score of user satisfaction increased dramatically when implemented, indicating that users were several times more appreciative of the increase in transparency and ethical considerations. Such growth has been a crucial factor in the subsequent adoption of RA platforms by increasing the trust of its users [10].

Tables and Graphs Illustrating Key Findings

Compliance Adherence Rates

The below Table 8.4 depicts compliance adherence rates.

Table 8.4 Compliance Adherence Rates.

Period	**Compliance Adherence Rate (%)**
Pre-Implementation	75
Post-Implementation	95

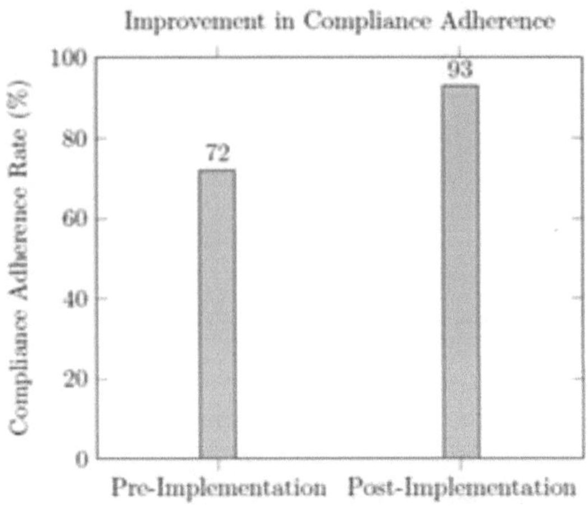

Figure 8.2 Compliance Adherence Rates Before and After Implementation.

The above Figure 8.2 shows the Compliance adherence rates before and after implementation.

Incident Reports

The below Table 8.5 indicates the incident reports before and after implementation.

Table 8.5 Incident Reports.

Period	**Incident Reports (Quarterly)**
Pre-Implementation	20
Post-Implementation	5

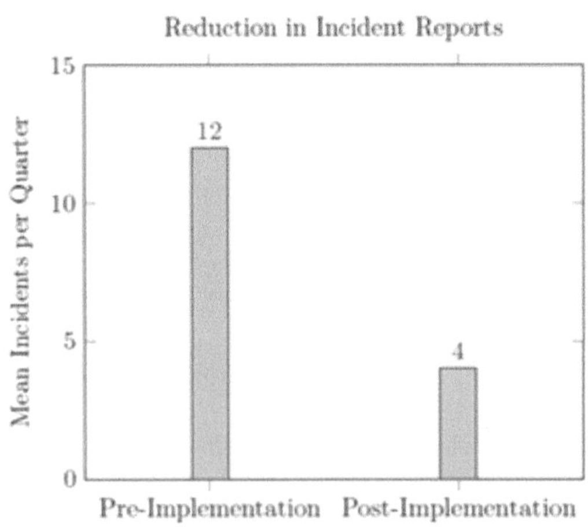

Figure 8.3 Incident Reports Before and After Implementation.

The above Figure 8.3 shows the reduction in incident reports before and after implementation.

User Satisfaction Scores

The below Table 8.6 depicts the user satisfaction scores. The below Figure 8.4 represents the user satisfaction scores before and after implementation.

The fine evaluation of the outcomes of the implemented changes emphasizes the extremely positive effect that the advertisement's proposed compliance and ethical framework had on InvestAI. All these enhancements in compliance adherence, a decrease in incident reports, and increased user satisfaction are indicators of the effectiveness of the framework. Such findings give a good argument for the wider application of such frameworks in the financial services market to provide strong compliance, increased ethical behaviour, and greater user confidence.

Table 8.6 User Satisfaction Scores.

Period	**User Satisfaction Score (Mean)**
Pre-Implementation	3.5
Post-Implementation	4.7

Figure 8.4 User Satisfaction Scores Before and After Implementation.

6. Regulatory and Ethical Implications

6.1 Consequences of New Frame: Regulatory

Several regulatory considerations could be identified in the introduction of the new compliance and ethical framework of the Robo-Advisors (RAs), such as InvestAI. The framework addresses the most pressing regulatory issues by enhancing the clarity of algorithms, security, and ethical adherence of data processing, and sets a new degree of compliance level within the financial services industry.

A key regulatory implication is that there can be better oversight and enforcement. Regulators will be able to audit and monitor the speed of RA operations more easily through increased transparency. The ability is essential in the context of making sure that RAs operate within the requirements of the current financial-related rules and regulations in the field, including the Markets in Financial Instruments Directive (MiFID II) within the European Union, which incorporates the rule of transparency and investor protection [13]. Besides, the automated compliance monitoring tools contained in the framework enable both real-time monitoring of regulatory changes and timely adjustments to algorithms, which virtually eliminates risks of non-compliance [4].

The other important implication is the possibility of harmonizing the regulatory standards around the world. The values instilled in the framework can serve as a guide for creating international regulatory standards, such as algorithmic transparency and continuous compliance monitoring. Such harmonization can allow cross-border investment operations and guarantee that RAs are regulated with equal expectations all over the world [1].

6.2 Ethics and Conflicts of Interests

The deployment and operation of RAs is an issue that must first consider ethical concerns. The focus on ethical standards that the new framework has and frequent audits can alleviate multiple ethical issues, such as the existence of algorithmic bias and the denial of conflicts of interest.

Having said that, (a) Algorithmic Bias: The biases that may arise in the non-human decision-making process of AI are among the most urgent ethical issues to discuss. Discriminatory outcomes may be based on the data used in training algorithms or on the algorithms themselves, creating bias. In the framework, this is mitigated by adopting the practice of periodic ethical audits, as well as making sure that the information and algorithms used support the principles of fairness and non-discrimination [8].

Conflicts of interest--RAs will have to negotiate conflicts of interest to remain trustworthy and have integrity. As an example, it is tempting to offer financial products that are rewarding to the RA or its partners, other than the investor. Such practices are clearly forbidden by the ethical orders given in the framework, and it also requires various recommendations and transactions to be transparent [9]. Moreover, to certify its connection to these ethical protocols, the company also regularly checks whether its compliance has any flaws [6].

6.3 Guidelines to Regulatory Bodies

Based on the results of this research, it can be suggested that the following recommendations should be taken into consideration by regulatory institutions in the framework of improving the monitoring and control of RAs:

i. **Act Decree Algorithmic Transparency:** Regulators must mandate RAs to be transparent in documenting and disclosing a step-by-step account of their algorithmic decision-making process. This involves the source and assumption data, as well as the reasoning in the algorithms. This transparency, in addition to facilitating regulation, increases investor confidence [13].

ii. **Introduce Continuous Compliance Monitoring:** Regulatory experts must promote the use of automatic instruments that accommodate compliance checking in real time. Such tools are able to monitor any changes in regulation and make sure that RAs adapt their processes accordingly as quickly as possible [4].

iii. **Enforce the Ethics:** Regulators need to set up clear ethical standards of RAs addressing concerns like algorithmic bias and conflicts of interest. Routine ethical reviews ought to be enforced to make sure that these guidelines are followed [9].

iv. **Encourage Data Security Best Practices:** Regulatory organizations should encourage data security best practices because such information is sensitive. These are superior encryption algorithms, strong data security mechanisms, and sound access control systems [12].

v. **Promote International Harmonization:** The global regulatory agencies are encouraged to respect each other in order to make cross-border investments via standardized regulations in the oversight of RA. Principles described in this framework can be used as a foundation of such harmonized standards [1].

vi. **Promote Invention by Use of a Sandbox:** Regulatory authorities should decide to take up some form of a sandbox to enable RAs to experiment with new compliance and ethical frameworks within a softly regulated environment. In this way, innovation can be conducted and new developments made; however, they should be in compliance and ethically correct [5].

By basing the regulation of RAs around these recommendations, the regulatory agencies can implement better control and management of such platforms and make such a system transparent, safe, and morally sound. Such precautions will assist in safeguarding the investor and preserving the level of the financial market as it grows in the dimension of using high-tech technologies.

7. Future Directions and Recommendations

7.1 Research Bases

The strategies used by Robo-Advisors (RA), such as InvestAI, to integrate a wholesome compliance and ethical framework, are a considerable move to increase transparency, data security, and ethics. Nonetheless, there are a number of areas that need to be examined further in order to continue narrowing down and redesigning this framework.

The analysis of more advanced approaches to ensuring algorithmic accountability should be conducted in the future. This involves creating robust methods of auditing AI models in order to identify and fix biases, creating tools that help one gain better insights into the decision-making of these algorithms [8].

i. **User behavioral analysis:** It is important to learn the user behavior and how they work with the recommendations of RA. Future research can examine the interaction between users and RAs, and it is possible to consider the differences in the perceptions and feedback of various population groups regarding the automated financial advice. This may assist in the customisation of RAs to user needs and increasing user trust [9].

ii. **Influence of Real-Time Data:** The use of real-time data in decision-making activities should be further discussed. Predicted future studies are a possibility to study how including real-time sources of data, like social media sentiment and economic indicators, can enhance the precision of the RA recommendation and their correlation [3].

iii. **Cross-Border Regulatory Compliance:** Since financial markets are going global, the need to examine issues and solutions to cross-border regulatory compliance is necessary. The study may be aimed at aligning regulatory norms and frameworks that may be flexible to meet various regulatory regimes with ease [1].

7.2 Learning to Cope with a Changing Regulatory Environment

The financial regulatory landscape changes continuously; therefore, regulatory frameworks of RAs should be able to keep up with the pace. The consideration of the proposed framework is positive as far as its modular design is concerned, although some additional measures can make the final product solid.

a) **Participation in Regulatory Sandboxes:** RAs may consider involving themselves in sandbox regulatory approaches that can enable them to improve and test the frameworks. It is also the process of experimenting with new technologies and methods of compliance with the oversight of regulatory institutions, which leads to innovation and maintains standards of compliance [5].

b) **Continuous Monitoring and Updating:** An important aspect is to install a system that continuously monitors changes in the regulation. This system ought to have automated notifications and updates, so that the RA framework will always be in accordance with the actual regulations. The compliance monitoring tools and ethical guidelines should be regularly updated in order to respond to new regulatory demands in a timely manner [4].

c) **Liaison with Regulatory Bodies:** Active liaison with the regulatory bodies will assist in forecasting the regulatory alterations and consequently be ready to utilize them. Such collaboration would also allow supplying quality feedback to regulators, which would be used to frame the next regulation that is not only effective but also viable to implement by RAs [13].

7.3 Scale-up Strategies in Implementation at Financial Institutions

Scoping the implementation of the compliance and ethical framework across different financial institutions would necessitate a tactical plan and a tactical implementation process. The following are some of the ways that can be employed to make this process easier:

a) **Standardization and Customization:** It is essential to develop a standardized form of the framework and render it easily adaptable in different financial institutions to fit their unique requirements. This balance between standardization and customization will provide the framework with the possibility of being both widely available and able to cover the needs of individual institutions.

b) **Training and Education Programs:** It is necessary to implement major education programs involving human resources in a variety of financial institutions. Such programs ought to focus on the principles of the framework, including the transparency of the algorithm, protection of data, and the principles of ethics. Continuous educational activities may lead to the maintenance of high compliance and ethical practices [9].

c) **Technological Integration:** The use of cloud-based solutions may help to roll out and scale the framework. The implementation of the new standards by institutions of any size can become more feasible because data on cloud platforms facilitates the scalable infrastructure and service, thus supporting the implemented framework [12].

d) **Pilot Programs:** By carrying out pilot programs across various financial institutions, it might be possible to identify possible barriers and improve the structure before full-scale deployment. They are pilot programs that provide priceless insights and enable changes to be made according to the practical experiences [8].

e) **Feedback Mechanisms:** It is vital to come up with fully functional feedback mechanisms to enable steady growth and development. It is suggested to involve financial institutions deeply in the process of giving feedback concerning the practical use and overall work of the framework, which can be used to conduct refinements in the future [3].

The envisioned compliance and ethical framework of Robo-Advisors like InvestAI has shown significant potential in increasing the regulatory compliance factor and ethical standards. Nevertheless, this could only be done through constant research and evolution so that it could always be relevant and effective. Future research can build on the framework by addressing algorithmic accountability concerns, user behavior, real-time data integration, and cross-border compliance. To ensure the strength of the framework, one should respond to the changes in regulatory environments by engaging in regulatory sandboxes, ongoing monitoring of the environment, and establishing a proactive dialogue with regulators. Finally, scaling should be implemented through strategic policies, training, technological integration, pilot programs, and the development of feedback mechanisms, which will promote mass adoption in the financial sphere.

8. Conclusion

The following paper proposes an extensive model of integrating compliance and ethical aspects into the Robo-Advisor (RA) of InvestAI. Application of this framework has led to radical advances in regulatory compliance, algorithmic transparency, data safety, and morality. The most important findings of the study represent that the adherence rate to compliance grew significantly, compliance-related incidents decreased significantly, and the satisfaction scores also increased. These findings prove the efficiency of the framework when dealing with both regulatory and ethical issues during the functioning of RAs [4].

Moreover, the research supplements existing literature by offering a comprehensive, data-driven analysis of implementation obstacles and solutions within the concept of applying that framework in a real-world RA system environment. A combination of full-scale AI, machine learning, and Natural Language Processing (NLP) solutions was proven to allow increasing the operative efficiency of RAs and the rate of user interactions, as well as guaranteeing sound compliance and moral practices [9, 8].

It cannot be overstated how crucial it is to incorporate compliance and ethics into the Robo-Advisors (RAs). With the financial industry leaning towards becoming more digital, there is an expanded need to use automated systems in terms of getting investment advice and portfolio management. By making sure that these systems operate within a strong compliance and ethical program, it is essential to keep investor confidence and ensure the integrity of the market [13].

By incorporating the spirit of transparency, accountability, and fairness into the RAs-driving algorithms, the suggested framework not only increases regulatory compliance, but also resolves the ethical issues at a higher level. These two emphases on compliance and ethics, in turn, help RAs deliver impartial, transparent, and safe financial guidance and, in such a way, protect the interests of investors and encourage sound market practices [4].

The involvement of compliance and ethics in the RA platforms extends beyond the individual investors into the entire financial industry. Among the main benefits of introducing such frameworks, one can also cite that the subsequent rise in transparency and accountability in the industry can act as an example to follow by other financial institutions in the country. That would help to put an end to the trend of irresponsible and unethical financial advisory services and establish a culture of trust and honesty [9].

Moreover, the results of the harmonization of regulatory standards that can be achieved with the help of such frameworks as the one suggested in the given study might boost cross-border investment activities. This will allow the global financial system to be more efficient and more integrated since financial institutions can better operate in foreign markets as the regulatory requirements they are subjected to are consistent worldwide [1].

The findings of the research study can be implemented in policymaking and regulatory changes in establishing more effective oversight systems based on automated financial services. This prospective orientation toward regulation can then assist in reducing the possible risks that accompany technology development in the financial sector, balancing it between innovation and protection of investors.

In a nutshell, compliance and ethical standards must be applied to the regulatory architecture in order to promote the responsible evolution of the financial industry through technological means. The proposed framework will enhance the integrity of the functions of regulatory architectures and improve the transparency and accountability of the financial ecosystem, besides being more reliable due to exhaustive regulation and ethical oversight across the entire framework.

Acknowledgements

Sincere thanks to Fr. Lijo for the constant support and guidance. Also, thanks to Mr. Ajay Garg for supporting the initial research idea, and Dr. Vivek Joshi for conceptualization and conducting a pilot study.

References

[1] Ahern, D. (2021). Regulatory Lag, Regulatory Friction, and Regulatory Transition as FinTech Disenablers: Calibrating an EU Response to the Regulatory Sandbox Phenomenon. Eur Bus Organ Law Rev. 22(1): 45–78. https://doi.org/10.1007/s40804-021-00217-z.

[2] Back, C., Morana, S. and Spann, M. (2023). When do robo-advisors make us better investors? The Impact of Social Design Elements on Investor Behavior. J. Behav Exp. Econ. 103: 101984. https://doi.org/10.1016/j.socec.2023.101984.

[3] Beketov, M., Lehmann, K. and Wittke, M. (2018). Robo Advisors: Quantitative methods inside the robots. J. Asset Manag. 19(6): 363–370. https://doi.org/10.1057/s41260-018-0092-9.

[4] Fein, M.L. (2015). Robo-Advisors: A Closer Look. SSRN Electron J. https://doi.org/10.2139/ssrn.2658701.

[5] Gregor, S. and Hevner, A.R. (2013). Positioning and presenting design science research for maximum impact. MIS Q. 37(2): 337–355. https://doi.org/10.25300/misq/2013/37.2.01.

[6] Gurrea-Martínez, A. and Wan, W.Y. (2021). The promises and perils of robo-advisers: challenges and regulatory approaches. SSRN Electron J. https://doi.org/10.2139/ssrn.3931448.

[7] Imerman, M.B. and Fabozzi, F.J. (2020). Cashing in on innovation: A taxonomy of FinTech. J. Asset Manag. 21(3): 167–177. https://doi.org/10.1057/s41260-020-00163-4.

[8] Jung, D., Dorner, V., Glaser, F. and Morana, S. (2018). Robo-Advisory: Opportunities and risks for the future of financial advisory. SSRN Electron J. https://doi.org/10.2139/ssrn.2970084.

[9] Lui, A. and Lamb, G.W. (2018). Artificial intelligence and augmented intelligence collaboration: regaining trust and confidence in the financial sector. Inf Commun Technol Law. 27(3): 267–283. https://doi.org/10.1080/13600834.2018.1488659.

[10] Phoon, K. and Koh, F. (2017). Robo-advisors and wealth management. J. Altern Invest. 20(3): 79–94. https://doi.org/10.3905/jai.2018.20.3.079.

[11] Seiler, V. and Fanenbruck, K.M. (2021). Acceptance of digital investment solutions: the case of robo advisory in Germany. Res. Int. Bus Financ. 58: 101490. https://doi.org/10.1016/j.ribaf.2021.101490.

[12] Strzelczyk, F. (2017). Cyber Risk and Cybersecurity: A systematic review of data availability. Geneva Pap Risk Insur Issues Pract. 47(3): 391–415. https://doi.org/10.1057/s41288-022-00266-6.

[13] European Parliament & Council of the EU. Markets in Financial Instruments Directive (MiFID II). Off J Eur Union. 2017. Retrieved from https://eur-lex.europa.eu/legal-content/EN/TXT/?uri=CELEX%3A32014L0065.

[14] Sanz Bayón, P. (2018). A legal framework for robo-advisors. Soc. Sci. Res. Netw. https://papers.ssrn.com/sol3/papers.cfm?abstract_id=3226644.

[15] Sumida, M. (2020). Robo-advisors and the legal duties of their providers. Soc. Sci. Res. Netw. https://ssrn.com/abstract=3746899.

Chapter 9

Blockchain-Powered Employee Life Cycle Management

A Framework Analysis For Transparency And Security

Sharad Gupta and *Mahesh Kumar T.**

1. Introduction

Employee Life Cycle Management (ELCM) is the chief stages of how an employee progresses from being recruited, onboarded, and working, followed by performance management, as well as offboarding. The presence of this phenomenon is very critical as it forms a crucial model for the optimization and integration of the concepts of human resource functions with the organizational goals [6]. ELCM contributes effectively to employees' satisfaction, engagement, and retention as they have a direct effect on productivity and organizational success [3].

Traditional ELCM systems, however, are inefficient, fragmented, lack transparency, and expose vulnerabilities in their security [2]. Most organizations have difficulty when dealing with sensitive employee data, complying with privacy regulations, and providing an easy experience for employees across the employee lifecycle [7]. There is no integration among several HR functions, resulting in scattered data management, making decision-making and process improvisation difficult [1].

Christ University, India.
Email: sharad.gupta@christuniversity.in
* Corresponding author: maheshkumart@christuniversity.in

DOI: 10.1201/9781003635178-12

The issue with traditional ELCM systems addressed above makes blockchain technology an interesting solution. Decentralization, immutability, transparency, and security—these core features of BTC fit well with the current needs of HR management [3]. Data silos can be eliminated by blockchain who's unified tamper-proof ledger for all employee-related transactions can improve transparency and trust [7]. Furthermore, smart contracts can resolve repetitive processes like payroll processing, benefits administration, company compliance verifications, and so on, and this resolves challenges related to inefficiencies [4]. However, limited adoption of the blockchain in ELCM has so far been observed. The majority of existing studies are isolated, functioning with [3] and not a holistic lifecycle approach. As a result, the inclusion of blockchain integration throughout each ELCM stage constitutes a major grey literature [2].

2. Objectives of the Study

The purpose of this paper is to cover the points identified earlier by:

- Building a complete KM system with advanced blockchain technology for ELCM.
- Showing how blockchain can increase transparency, provide greater security, and improve performance from start to finish in a person's employment.
- Applying the design-by-case study and physical prototype to confirm that it can be used for real challenges.

The research aims to make the ELCM process more secure and open. The goal of the proposed framework is to take advantage of blockchain's key abilities. It is important to stop any breaches or unauthorized access to staff data [5]. Do not tolerate inefficiency; use automation and integrate processes to fix them. Ensure employees and organizations feel trustworthy by securely storing and verifying all lifecycle events [7].

3. Literature Review

Human Resource Management (HRM) is being transformed by blockchain technology, which helps deal with inefficiencies and offers secure, transparent, and decentralized solutions. Blockchain, unlike conventional HR technology, stores information in a way that is hard to change and provides access to everyone, removing the risks of both data tampering and unauthorized entry. This review investigates the ways in which blockchain can improve employee records, simplify human resource processes, and build trust between staff and management.

Because blockchain cannot be altered without agreement, records of employees are secure and can always be checked. Since data must be correct in HRM, blockchain can stop problems from cropping up when data in key areas like payroll, benefits, or performance tracking is corrupted or illegally changed [7]. For instance, payroll mistakes occur often in organizations because of miscommunication or poor handling of data. With smart contracts, blockchain-based payroll systems ensure that all payments are accurate and employees receive them on time. All networks record each transaction on the blockchain, letting all stakeholders have clear visibility into all activities while keeping their private financial information safe. Another major benefit of blockchain is how clear it makes things. Both employees and employers view reliable records of employment history, certificates, and assessment reports. As a result, both employees and employers trust the data, as candidates or employees cannot lie about their history without being caught since the firm verifies data independently [6].

Digital Identities on the Blockchain

Companies spend a lot of time during recruitment, checking everything from someone's degrees and diplomas to all the places they have worked. These traditional methods of verification are both slow and expensive, and easily subject to fraud. If someone's credentials are on a blockchain, they are instantly available for check and do not require verification from the organization. Because they have safe, verified credentials on the blockchain, employees using digital identity systems can easily present them to employers when seeking jobs. Without intermediaries, the risk of someone receiving fake credentials is minimized [2]. They can also be sure the qualifications are secure, allowing them to hire more quickly and spend less on administration. For example, Workday and SAP are investigating using blockchain to improve the way recruitment is handled. These systems make it easier for people to remain compliant with data protection rules by enabling them to control who sees their personal information [5].

Automated processes are supported via smart contracts, which trigger themselves when their defined terms are satisfied. These contracts are smart for HR tasks that happen frequently or must follow strict company rules. In payment systems for employees called payroll, smart contracts ensure that individuals get their earnings when the criteria, such as their performance or attendance, are fulfilled. As a result, paperwork for payments is decreased, mistakes are avoided, and payments are sent promptly [4]. Benefits management also relies on smart contracts to provide useful solutions. Traditionally, benefits administration is slow and less efficient because it depends on paperwork and requires manual checks. Thanks to blockchain, employees get their benefits without having to wait for a long time. So, with necessary documents verified on the blockchain, healthcare claims can be handled by smart contracts and instantly approved.

Engaging staff positively boosts their performance and reduces the chance of them leaving. A new approach to employee rewards and recognition has arrived with blockchain through tokenization. Groups can introduce blockchain tokens to reward staff for meeting their work goals or finishing training. Staff members can use their earned tokens to get tangible benefits, which makes recognition fair and easy to understand [3]. Using tokens guarantees that rewards are listed openly and fairly, helping to avoid unfairness when deciding what is recognized. The Cartan app allows employees to confirm their rewards transparently using the blockchain.

Figure 9.1 in VOSviewer shows all the themes and topics in our analysis and the ways they link together in relation to Blockchain-Enabled ELCM. This cluster

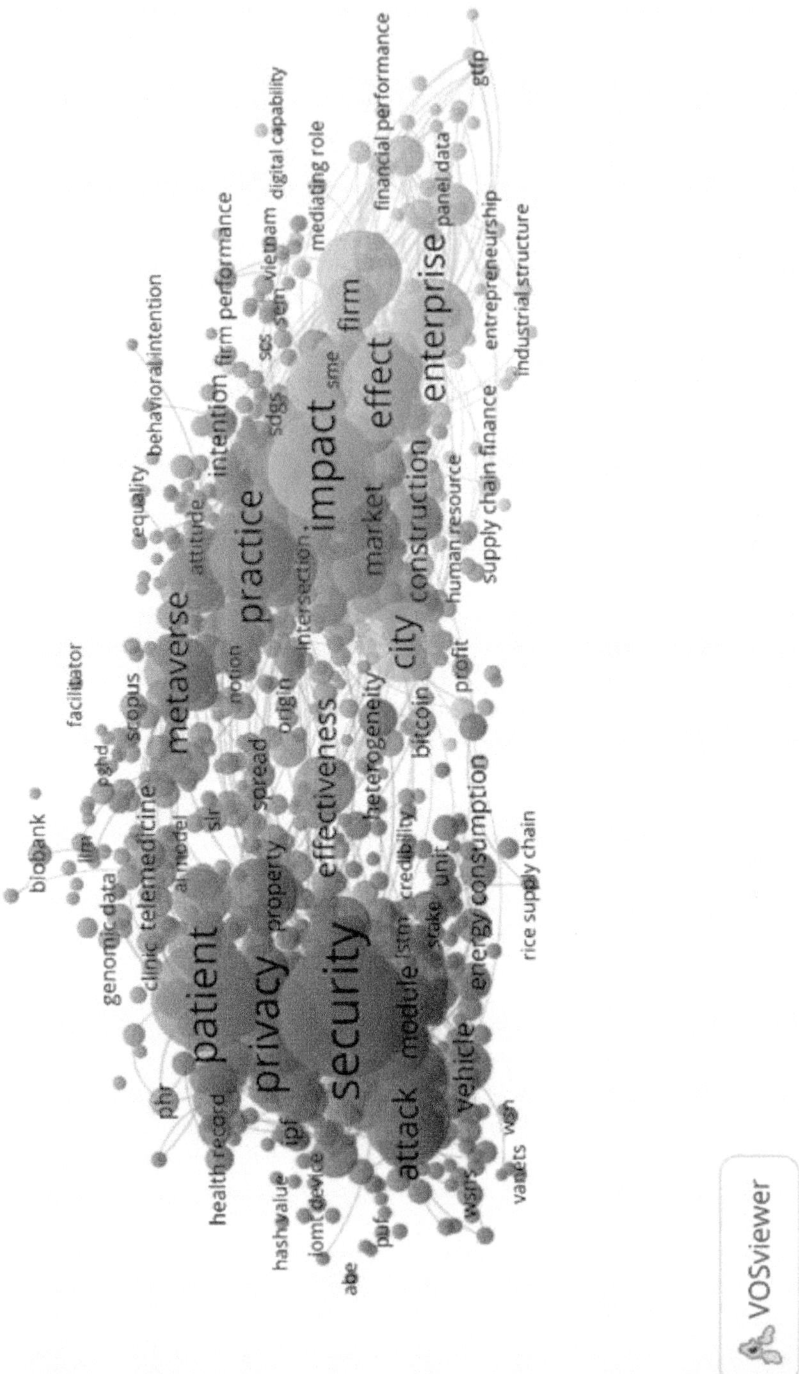

Figure 9.1 Text analysis-based thematic clustering for blockchain-enabled employee lifecycle management (ELCM).

contains topics that deal with security, privacy, and attack, reiterating the need to protect important employee information and patch security holes in digital HR systems. The blue cluster addresses topics such as the metaverse, telemedicine, and genomic data, meaning that progressive digital solutions help with employee management and organizational workflows. It appears from the green cluster that blockchain technology can support changes in the workforce in urban settings and contribute to the growth of sustainable enterprises. The yellow cluster addresses organizational issues, mentioning important terms "enterprise," "firm," and "effect," to show the effect of blockchain on improving efficiency and financial results for enterprises. By illustrating the diagram, we find that the area of Blockchain-Enabled ELCM is informed by many disciplines, mainly data protection, technological advancements, and organizational shifts.

Even with all its advantages, there are still challenges stopping HRM from adopting blockchain. Fewer people can use the network because it often gets slow and costly when there are many transactions to process. With popularity, public blockchains like Ethereum can get so busy that they slow down HR operations [5].

Working with existing HR systems is a further difficulty. In many cases, older systems in organizations neither integrate with blockchain nor can they be easily updated. At this moment, there is still a lot of change happening in blockchain regulations when applied to HRM [2].

Through blockchain, companies can boost the accuracy of their HR records, automate tasks, and ensure better transparency. Using blockchain in payroll, digital identity, and employee awards helps solve old problems in how HR is handled today. Even so, it is important to deal with problems like scale and matching it with current systems if we want to get the most out of it. First, future studies must target overcoming these challenges and then develop whole frameworks for using blockchain in HRM.

Traditional ELCM encounters many problems because of scattered systems, limited transparency, and risks in managing private employee details. Now that people work from many different locations, it is more difficult for companies to follow the rules in each place [6]. When HR data is in different, isolated systems, there are problems sharing employee information for recruitment, payroll, and performance management purposes. The fragmentation results in inaccurate data, sets back important choices, and causes workloads for managers to increase.

Security holes make the issues even more complex. Since traditional HR systems mainly store information in one place, they can easily be vulnerable to theft. Because of the personal and professional information kept in these systems, they become important targets for cyber criminals. An instance of a breach might affect employee trust and, at the same time, lead to severe consequences from legal and financial responsibilities [7].

Safeguarding employee data is now more complicated, thanks to new data protection laws such as the General Data Protection Regulation (GDPR). Sensitive worker data must be safe in storage and can be used only when the employee agrees, but this is hard to ensure using old systems. Failure to follow regulations leads to large fines and a damaged reputation, which has made it urgent to invest in systems offering strong security and clarity [1].

Blockchain may be used in ELCM

Blockchain can provide new answers to the problems and weaknesses in traditional ELCM. All employee data is kept secure on the blockchain because it cannot be changed without permission. By using this approach, we do not depend on a single system, reducing the risk of data being attacked and keeping our data reliable through each employee's time with us [2].

One of the main advantages of blockchain is that smart contracts can handle routine work, such as payroll, benefits, and compliance monitoring. A smart contract could set up payments to happen as soon as required conditions, such as showing up at work every day, are met. In turn, it relieves administrators, reduces the risk of mistakes, and speeds up completing HR work [4].

With blockchain, employers and employees can both monitor performance in real time, so all involved remain honest and responsible. They can review their progress on a secure system, meaning their progress and rewards depend on real data instead of personal preferences [7].

When data silos are broken, security is boosted, and key tasks are automated with blockchain, ELCM can undergo a major transformation to become more transparent, efficient, and secure.

Although there is strong potential for HRM with blockchain, most studies currently work on just single functions, particularly payroll or digital identity [5]. In addition, having a blockchain network work together with old programs is a main difficulty, as blockchain often has difficulty being compatible with other systems [3]. A further gap exists because sexual validation of blockchain in HR is lacking, with most models still at the conceptual level [7].

Experts explain that blockchain is uniquely placed to address the shortcomings and security risks found in ELCM. Yet, significant issues exist in actually adopting complete ELCM solutions and in proving them scientifically, so a complete blockchain framework is still needed.

Although HRM has seen positive results due to blockchain technology, existing research mostly examines limited subjects such as payroll, proving one's identity online, and providing benefits. These studies looked at only one segment, so there is still no holistic framework for using blockchain in all parts of the employee life cycle [5]. For key recruitment, onboarding, performance, training, and offboarding stages to be connected, companies need systems that work together, but current blockchain use mainly helps solve selected issues rather than tackling all lifecycle challenges.

Since blockchain has difficulty interoperating and scaling, it is not widely used in HRM. With the use of legacy infrastructure, it is hard for traditional HR systems to work with blockchain technology. Because legacy systems usually do not fit blockchain's decentralized model, they result in inefficiencies and incomplete implementation [3]. It is often hard for organizations with a high volume of transactions to handle changes in their environment. For example, the Ethereum blockchain, which supports smart contracts, may become clogged and charge high fees, which decreases its suitability for using HR applications by many companies [7].

A further challenge is that there is not much real-life research on how blockchain works in HR. Even though blockchain is found useful by theory and some small projects, large projects using it are still seldom researched. There is a shortage of data in existing blockchain models about their impact on organizational operations, how they make employees feel, or how laws are followed [2]. Conducting research is essential to measure the pros and to handle problems such as costs to onboard people, HR training, and how much employees learn about blockchain.

The literature does not focus enough on data privacy and following worldwide regulations such as the General Data Protection Regulation (GDPR). Since data on blockchain cannot be changed, it is hard to respect the “right to be forgotten” section of GDPR [1]. Because there are no secure strategies for meeting regulations and being transparent, businesses are still uncertain about blockchain in ELCM.

Also, user experience in blockchain is still a subject that has not received much attention. Most blockchain frameworks prioritize technology over ease of use. As a consequence, blockchain-based systems are challenging for HR to implement and use, thereby diminishing their overall benefits to the industry.

The papers point out that blockchain can solve problems, create more openness, and ensure safer data in ELCM. The technology’s automatic handling of similar jobs, safeguarding employee information, and streamlining recruitment and employee joining are key reasons why it is changing HRM [4]. But there are still important gaps, mainly in the complete use of blockchain to manage all aspects of the employee lifecycle. Research in this field is limited to specific apps, and important opportunities to collaborate, monitor operations, and view the entire lifecycle chain are not being realized [5].

Information on how blockchain improves organizational efficiency, the way employees feel, and compliance is mostly lacking. Lack of data may cause firms to be reluctant about using blockchain for ELCM, mainly due to its high starting costs and technical difficulties [2]. Moreover, challenges related to how systems connect, how many users they can support, and meeting regulations need to be handled to allow blockchain to be widely accepted digitally.

Figure 9.2 depicts the global research collaboration network in Block chain ELCM. Such gaps reveal that a complete blockchain-equipped ELCM framework should handle both technical and operational issues. The use of results from scientific research, user needs as the main design priority, and meeting rules from across the globe are all important for the framework to function well [6, 7]. Addressing these differences allows blockchain to make a real difference in ELCM by improving efficiency, security, and how employees believe in the HR department.

Traditional ELCM encounters many problems because of scattered systems, limited transparency, and risks in managing private employee details. Now that people work from many different locations, it is more difficult for companies to follow the rules in each place [6]. When HR data is in different, isolated systems,

Figure 9.2 Global Research Collaboration Network in Blockchain-Enabled Employee Lifecycle Management (ELCM).

there are problems sharing employee information for recruitment, payroll, and performance management purposes. The fragmentation results in inaccurate data, sets back important choices, and causes workloads for managers to increase.

Security holes make the issues even more complex. Since traditional HR systems store information in a single place, they can easily be vulnerable to theft. Because of the personal and professional information stored in these systems, they become important targets for cyber criminals. An instance of a breach might affect employee trust and, at the same time, lead to severe consequences from legal and financial responsibilities [7].

Safeguarding employee data is now more complicated, thanks to new data protection laws such as the General Data Protection Regulation (GDPR). Sensitive worker data must be safe in storage and can be used only when the employee agrees, but this is hard to ensure using old systems. Failure to follow regulations leads to big fines and a damaged reputation, which has made it urgent to invest in systems offering strong security and clarity [1].

4. Implementation and Validation

For prototype implementation, the ELCM framework was set up on Hyperledger Fabric because it has a modular design, can be used to create permissioned networks, and supports enterprise apps well. The way Hyperledger is built ensures that only those with access can use employee data to protect their information in HRM systems [7].

This tool includes several different components, all connected.

- The Distributed Ledger Module guarantees that all parts of employee data remain immutable from the time the employee is hired until they leave. Any transaction, such as employment or salary amendments, appears as a block on the ledger. Thanks to its distributed design, the ledger updates in real time, helping prevent errors in the data.
- Smart Contracts automate regular HR procedures. Attendance records and performance scores, set in advance, trigger the automatic processing of payroll. Such contracts are written in Chaincode, which is Hyperledger's own smart contract programming language [4].
- This module uses blockchain technology to verify employees' identities.

Externally accessed databases are accessed via APIs to check someone's education and work experience during the recruitment process. Using blockchain, each employee receives their own unique identity and safe, unchangeable document copies.

The interaction between the blockchain and traditional HR systems was improved through RESTful APIs. These APIs enable the blockchain system to integrate with reliable HR tools such as SAP HR and Workday. Furthermore, the framework design and deployment phase was much simpler due to Hyperledger Composer. Because the prototype was implemented in cloud computing, companies of all sizes can test ideas and ensure their applications are accessible. To prove the framework's worth, a case study was run using a simulation of a mid-sized multinational company. The processes, recruitment, payroll, performance management, and even offboarding were all moved to the blockchain.

- Selected candidates would be sent details for a test of their digital identity. By recording qualifications and experience on blockchain, the average time for verification fell from 5 days to just 1 hour.
- Smart contracts automatically process payroll for staff. With the system in place, every transfer was made exactly on time because it was triggered by set rules.
- Performance Management: The use of blockchain recorded KPIs for employees, so both employees and their managers could view current performance data anytime. It allowed the company to be more transparent about appraisals and promotions.
- When members left the system, their timely exit was safely and completely recorded by the blockchain.

The evaluation of the framework was based on using:

1. The time required to finish recruitment verification and carry out payroll tasks.
2. Clear communication: Feedback on how much employees can see and access their performance documentation.
3. Using RPA helps reduce expenses for administration and the mistakes associated with more traditional ways of working.

The evaluation found that administrative costs dropped by 40%, processes became 75% more efficient, and there was a major increase in employee trust as a result of improved transparency [3].

5. Results and Analysis

The ELCM framework developed using blockchain was better than traditional HR software in many ways.

Traditionally, checking the trustworthiness of a recruit is slow and can have errors because it relies on people checking the paper trail. Verification was cut down by more than 80%, and all issues related to fake credentials were taken away with the blockchain approach, [2] reported (2022).

Traditional payroll has problems with mistakes and delays caused by tasks that employees have to perform by hand. Thanks to blockchain payroll, there were no errors, and everything was processed instantly. Also, thanks to the blockchain, employees now have access to all immutable and checked performance records, while in traditional management practices, transparency was often compromised.

Key Performance Indicators are known as KPIs.

1. Data in the blockchain system is accurate to 99.9%, unlike traditional data which only achieves 95% accuracy [7].
2. During the launch, no security breaches came to light because data access was managed by permissioned nodes that used encryption protocols. Because traditional systems usually place all data in one central place, they experience more security breaches.
3. Users' Experience: The survey of imitation users found that 30% more employees trust the system and are more satisfied, due to the clear and speedy operation of blockchain.

6. Information on Scalability and Integration

Though there is high efficiency on the blockchain in a controlled situation, making the network work for millions of transactions is still an issue. Layer-2 solutions, such as Polygon, may be explored in future stages to deal with these issues [3].

APIs were successfully used to integrate with old HR systems, but additional research is important to make sure interoperability protocols can be applied easily.

By deploying and verifying the blockchain-based ELCM framework, it is clear that HR tasks will be reformed. According to the study, the new framework did better than previous systems when it came to efficiency, transparency, and security, confirming its suitability for today's HR challenges. Even so, scaling and long-term integration pose some issues that still need attention and development. By addressing these concerns, blockchain may soon be the basis of future ELCM systems.

Challenges and Limitations

Still, there are some obstacles to using blockchain technology for Employee Life Cycle Management (ELCM) despite its big potential. The main types of obstacles are insufficient technology, difficulties in convincing users, and problems related to ethics and rules. Solving these problems is vital to making the most of blockchain in HR.

Technical Challenges

1. Increasing the capacity of a blockchain and making transactions happen faster

   Blockchain technology needs to overcome its main technical limit, which relates to how many transactions it can process at once. Many transactions on Ethereum and similar public blockchains often cause congestion because of their inability to keep up with big volumes. If there are thousands of employees, the organization could have to deal with delays in everyday operations and a high number of transaction fees [7]. While Hyperledger Fabric and similar blockchains alleviate some limitations of scalability, they need substantial changes to computing equipment and software to perform tasks at a higher rate [3].

2. Working together with existing HR systems

   Blockchain is not planned into these HR systems because many organizations continue using them. It is usually tough and expensive to introduce blockchain because these systems have data formats that do not match the old infrastructure. On top of this, using blockchain with business applications such as SAP or Oracle HR calls for special middleware and extensive work on APIs, which makes everything more difficult [2].

Adoption Challenges

Organizational Resistance

1. Adopting blockchain in an organization is often challenged by its stakeholders. Both workers and human resource members may fear that blockchain is too hard to grasp and work with. Moreover, choosing to adopt blockchain is sometimes delayed by doubts about how new it will remain and the lasting value it might provide [6].
2. Blockchain's advanced nature makes it important for people working on it to have development, deployment, and maintenance skills. Still, a lack of skilled blockchain professionals creates a major problem. Managers need to either hold workshops or hire professionals, which takes longer and costs more [3].
3. To bring blockchain into a system, you have to purchase costly hardware, software, and train your team. It is important for organizations to set aside money for keeping blockchain systems maintained and updated, since these networks develop quickly. For SMEs, these costs can be too high, stopping many from adopting these technologies [4].

Problems Related to Ethics and Regulations

1. Legal blockchain solutions in ELCM have to be in line with the labor laws and rules across the world. Working toward GDPR compliance in Europe means companies need to handle employee data with great care. Blockchain's durable nature makes it hard to comply with the GDPR requirement that requests to erase data be followed [1].
2. Blockchain frameworks commonly ask employees to agree before their data can be stored. When there are no clear guidelines and honest exchanges, organizations may break privacy rules or lose their team's trust. All employees must know what blockchain means for them before giving consent [2].
3. Blockchain could be used in unethical ways in HR, and this is cause for worry. Blockchain's openness makes sure companies are accountable, though seeing every employee's data can cause stress or unfairness in the workplace. Appropriate behavioral guidelines are needed to prevent employees from facing this issue and to protect them [3].

ELCM could be transformed by blockchain, but achieving that transformation depends on overcoming these difficulties. Because of issues such as scalability and compatibility, networks like Layer-2 and middleware APIs are needed as advanced solutions. Because adoption can be hard due to organizational problems and budget limitations, it requires strong change management processes and evaluations of benefits versus expenses. In the end, ethical and legal rules require that blockchain systems are compliant, transparent, and support employees. Handling these difficulties will open the door for broader blockchain use in HR, which can improve how things happen, how agencies work, and their credibility.

7. Conclusion and Future Work

With the help of blockchain, managing employees has become simpler and more reliable. Since blockchain distributes data across multiple computers, employee data is now more secure from data breaches. Because blockchain keeps an unchangeable record, it is impossible for employee data to be tampered with, guaranteeing its accuracy at any time. As a result, stakeholders—employees, employers, and regulatory groups—feel more confident in each other.

Blockchain helps make HR systems much more transparent. Thanks to real-time data on performance, evaluations, and benefits, employees are more willing to trust their employer and less likely to have disputes. Blockchain enables employers to verify recruitment, payroll, and other compliance procedures easily, meaning they are always accountable and secure. Smart contracts take care of tasks such as payroll and benefits assignments automatically, which decreases both errors and operating problems [3]. Handling these challenges through the proposed framework, blockchain can help make employee management clearer, more secure, and more efficient. The process makes HR practices in modern companies more equal and reliable.

The ELCM framework, based on blockchain technology, offers innovative solutions for the practice of HR management.

1. Unlike most HR blockchain solutions that handle payroll or identity verification only, this framework uses blockchain to support every step in ELCM. All the steps, from hiring to ending someone's contract, gain from being secure, efficient, and clear [5].
2. Smart Contract Automation The framework introduces smart contracts that help handle repeated activities without human participation and guarantee accurate workflows. With the use of these contracts, the framework reduces delays and errors in payroll, benefits management, and compliance, doing better than traditional HR methods [4].
3. Blockchain's openness and tokenization help make the work environment more rewarding for employees. Workers can check their achievements and compensation as they happen, which helps hold everyone accountable and acknowledge their success [2].

4. The framework helps firms meet GDPR standards by supplying a safe and open system for managing their employees' details. Thanks to blockchain's unchangeable nature, it supports compliance and privacy by allowing access to data only by approved users [1].

8. Future Possible Research Themes

1. Scalability of blockchain is a central issue blocking its wider use in ELCM. When Ethereum and similar networks handle big volumes of transactions, their performance starts to suffer. Research projects can investigate integrating Polygon or other Layer-2s or using blockchains such as Hyperledger Fabric to support the scalability of major companies [3].
2. By combining Artificial Intelligence (AI) with blockchain, companies can find new ways to guide their HR decisions. As a result, AI tools can analyze blockchain data on worker performance to predict who might quit, offer individual development plans and ensure the right number of people are assigned to different positions. The main task should be to integrate AI analytics into blockchain while not compromising its usual transparent and secure features [7].
3. Multi-Stakeholder Collaboration: Successful ELCM needs to support interactions among employees, employers, government officials, and outside service suppliers. In the future, researchers could try to connect blockchain networks with the traditional systems common among stakeholders. Onboarding different parties, including data center operators, posting site owners, and network providers, can be made easier if standardized APIs and middleware are set up [2].
4. Blockchain's transparency and immutability in HRM reveal various ethical issues that must be explored. A main aim of research should be to form regulations that strike a balance between displaying information and respecting people's data, in accordance with GDPR and similar standards. In the future, researchers need to focus on how blockchain technology could cause ethical issues, such as more employee supervision and using permanent performance data for discrimination [1].

The ELCM framework, powered by blockchain deals with major problems in traditional HR systems by making everything clearer, safer, and more efficient for all employees. Though this framework shows progress, new research is needed to overcome issues such as how blockchain will be used in many areas, how it can communicate with existing systems, and what ethics are involved. When AI is used, and many parties work together on blockchain, HRM can change for the better, choosing efficient, fair, and employee-centered approaches.

References

[1] Cattermole, G. (2019). Developing the employee lifecycle to keep top talent. Strategic HR Review.
[2] Gladka, O., Fedorova, V. and Dohadailo, Y. (2022). Development of conceptual bases of the employee life cycle within an organization. Business: Theory and Practice.
[3] Gupta, S. and Sharma, R. (2024). EmploChain: A blueprint for blockchain-driven transformation in employee life cycle management. In the 2024 International Conference on Trends in Quantum Computing and Emerging Business Technologies.
[4] Hauschild, M., Kara, S. and Røpke, I. (2020). Absolute sustainability: Challenges to life cycle engineering. CIRP Annals-Manufacturing Technology 69: 533–553.
[5] Murali, B. and Kalpana, J. (2020). A study on employee life cycle management. Journal of Emerging Technologies and Innovative Research.
[6] Parkar, S. and Kshirsagar, D. (2024). A study on employee life cycle management. International Journal of Scientific Research in Engineering and Management.
[7] Xiang, H., Lu, J., Kosov, M., Volkova, M., Ponkratov, V., Masterov, A. et al. (2023). Sustainable development of employee lifecycle management in the age of global challenges: Evidence from China, Russia, and Indonesia. Sustainability.

Security and Data Protection

Chapter 10

Preserving Format, Protecting Data

A Multi-Field Encryption Technique

S. Vidhya

1. Introduction

Format-preserving encryption (FPE) is encryption at the field level for small fixed-length sensitive data inside databases, such as systems. It preserves the original data format, so that the information remains meaningful within the database environment, indexes, or other support structures. Since it maintains its format, all application queries and manipulations of this encrypted datum do not require major adjustments in the underlying structures or logic of applications. FPE, therefore, proves extremely useful for relatively short fields of small fixed-length data. Traditional encryption makes nonsense out of intelligible data; in contrast, FPE allows encoded data to be queried and scanned. That means an application can sort and filter, as well as index and search encrypted data, thus reading and analysing data efficiently without breaking security.

Michael Brightwell et al. [1] were the first to propose the FPE scheme. The major methods proposed were the prefix cypher, cycle-walking cypher, and generalized Feistel cypher. These were developed using the Rank-then-Encipher (RTE) technique with an imbalanced Feistel structure. RTE enables FPE to be applied to various formats as shown in Figure 10.1. The National Institute of Standards and Technology (NIST) has approved the algorithms BPS, VAES3, and FFX, with the relevant standards being FF1, FF2, and FF3. However, FF2 was excluded from the FFX modes as it failed to meet NIST's security criteria. In FF3-1, the tweak parameter was reduced

Associate Professor, Department of Computer Science, KG College of Arts and Science, Coimbatore, Tamil Nadu, India.
Email: vidhya.s@kgcas.com

DOI: 10.1201/9781003635178-14

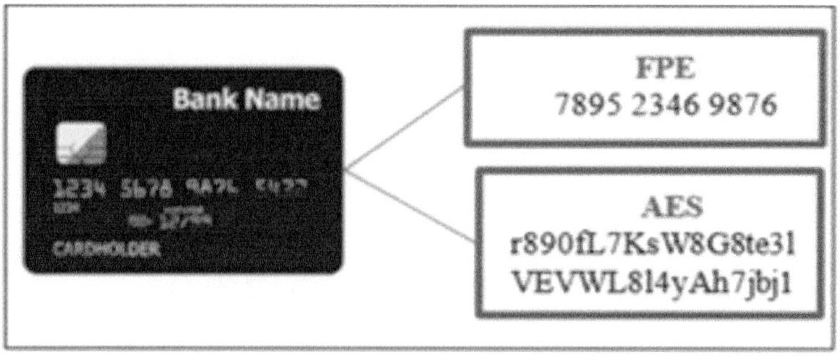

Figure 10.1 Credit Card Number encryption by Format-Preserving Encryption and AES.

to 56 bits from 56 bits. The current Format-Preserving Encryption standards are FF1 and FF3-1 [2].

2. NIST Approved Format-Preserving Encryption Algorithms

2.1 Cryptographic Transformation

FF1 and FF3-1 use block cipher encryption to preserve the plaintext format. By default, FF3-1 employs the AES block cipher in Counter (CTR) mode. While FF1 may use a different block cipher, the specification suggests using the Advanced Encryption specification (AES) as the default [3].

The cryptographic functions of the FF1 and FF3-1 schemes, as well as the associated parameters, are shown in the following notations.

Encrypt (Plaintext, Security-key, Tweak) = Ciphertext
Decrypt (Ciphertext, Security-key, Tweak) = Plaintext

Decryption is the reverse of the encryption function, which is also represented by

Decrypt (Encrypt (Plaintext, Security-key, Tweak), Security-key, Tweak) = Plaintext

2.2 Feistel Structure

The Feistel network architecture as shown in Figure 10.2, which is the architecture used in block ciphers, provides the basis for the FF1 and FF3-1 modes. In both FF-1 and FF3-1, the data block is divided into two segments. The round function is applied to two segments separately and merged. This process is followed for all the rounds—the generic process of the feistel structure.

Division of Data: The input data block is split into two halves, named the left half (L) and the right half (R).

Round Function: Round key and the right segment of R are the inputs for the round function, which yields the output (R').

Exchange and Input: The right and left parts are switched at the conclusion of each round. The left half becomes right, and the new right (R') becomes left.

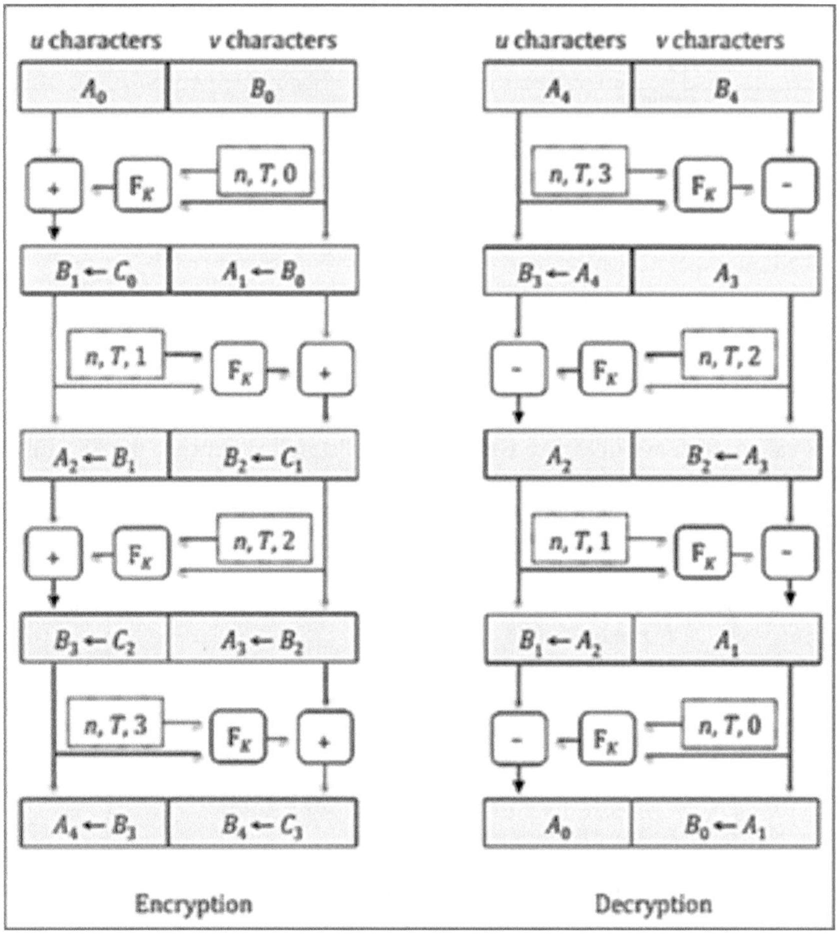

Figure 10.2 The FF1 and FF3-1 design based on the Feistel Structure.

Last Round: Without switching the halves, a final round is played after a certain number of rounds.

Results: The left segment (L) and the right segment (R) are concatenated to form the ciphertext, which is the result of the Feistel network.

2.3 AES-FF1

The FF1 algorithm can handle several data types, Table 10.1 represents the various parameters including strings and integers, and is designed to function on binary strings of any length [3].

The encryption process described above splits the plaintext X into two parts, A and B. For the next encryption step, the internal function uses the length of B (b), along with d and P (the first block). In addition, Q is produced from B using the tweak value (T) and the round number (i). The first two blocks, P and Q, are then

Table 10.1 Parameters in FF1.

Parameter	Value Range	Description
minlen	2	Minimum length of input
maPTlen	232	Maximum length of input
radix	2 to 2^{16}	Radix (base) range
tweak T	0	Optional tweak value
round	10	Number of rounds

used to create R, while S is obtained by encryption. The decimal strings that come from S and A are concatenated with the modulo the radix value (m). Ten times during this conversion procedure, the original length m radix frequency value is included.

The encrypted value is represented by the resultant value that links the final A and B values as shown in Figure 10.3. The only distinction between the FF1.Encrypt and FF1.Decrypt algorithms are at Step 6: In Step 6vi, (1) the indices are arranged in reverse order, (2) A and B's values are swapped, and (3) modular subtraction is utilized in place of modular addition as shown in Figure 10.4.

```
1. Let u = ⌊n/2⌋; v = n – u.
2. Let A = X[1..u]; B = X[u + 1..n].
3. Let b = ⌈ ⌈v·LOG(radix)⌉/8 ⌉.
4. Let d = 4⌈b/4⌉ + 4.
5. Let P = [1]1 || [2]1 || [1]1 || [radix]X
   3 || [10]1 || [u mod 256]1 || [n] 4 || [t]4
6. For i from 0 to 9:
   i. Let Q = T || [0](−t−b−1) mod 16 || [i]
     1 || [NUMradix(B)]b
   ii. Let R = PRF(P || Q).
   iii. Let S be the first d bytes of the following
      string of ⌈d/16⌉ blocks:
       R || CIPHK (R ⊕ [1]16) || CIPHK (R ⊕
       [2]16) ... CIPHK (R ⊕ [⌈d/16⌉–1]16).
   iv. Let y = NUM(S).
   v. If i is even, let m = u; else, let m = v.
   vi. Let c = (NUMradix (A)+y) mod radix m .
   vii. Let C = STR m radix (c).
   viii. Let A = B.
   ix. Let B = C.
7. Return A || B
```

Figure 10.3 FF3-1 Encryption algorithm.

1. Let u = ⌈n/2⌉; v = n – u.
2. Let A = X[1..u]; B = X[u + 1..n].
3. Let TL = T[0..27] and TR = T[28..55]
4. For i from 7 to 0:
   i. If i is even, let m = u and W = TR, else let m = v and W = TL.
   ii. P = W ⊕ [i] 4 || [NUMradix (REV(A))]12.
   iii. Let S = REVB(CIPHREVB(K) REVB(P)).
   iv. Let y = NUM(S).
   v. Let c = (NUMradix (REV(B))–y) mod radix m .
   vi. Let C = REV(STRmradix (c)).
   vii. Let B = A.
   viii. Let A = C.
5. Return A || B.

Figure 10.4 FF3-1 Decryption algorithm.

2.4 AES-FF3-1

Small domains are mostly served by FF3-1. To expand the domain space, it also uses Tweak as shown in Table 10.2. To achieve high entropy in each half, the adjustments should be generated at random [4].

An algorithm similar to FF1, the plaintext X is divided into two segments, A and B. The 54-bit random variable is used. The combination of T[0..27] and 04 is used to create TL, and the combination of T[32..55], T[28..31], and 04 is used to create TR.

The i represents the round number. W takes the value of TR, and m is allocated to u if i is an even integer. On the other hand, W is set to TL and m is allocated as v if i is odd. Using W and i, a four-byte value is created.

P is constructed by merging the two radix-notation numbers, B and P, which are created by reversing B, and 12 bytes of data are created by padding it with leading zeroes. S and A are used to create C in a manner comparable to FF1. FF3-1. It takes eight rounds of the round function to get the value of C. The final values of A and B are combined to produce the encrypted text.

Table 10.2 Parameters in FF3-1.

Parameter	**Value Range**	**Description**
minlen	2	Minimum length of input
maPTlen	$2 \times \lfloor \text{logradix}(2^{96}) \rfloor$	Maximum length of input
radix	$2..2^{16}$	Radix (base) range
tweak T	56 bits	Optional tweak value
round	8	Number of rounds

3. Proposed Algorithm

Format-preserving Encryption standards FF-1 and FF3-1, approved by NIST, are deterministic, in which the same plaintext is always encoded into the same ciphertext. In some circumstances, deterministic encryption may allow information to leak out, particularly when encrypting repeated or structured data. The encrypted text may be subject to frequency or pattern analysis due to the absence of probabilistic encryption [5]. By interleaving multiple fields to construct the plaintext, the suggested approach enhances the complexity of the traditional FF3-1 algorithm.

3.1 Secured Format Preserving Encryption

The suggested technique adds an extra degree of protection and adaptability to the encryption scheme, drawing inspiration from FF3-1. To test the suggested algorithm, the credit card number, expiration month, and date are encrypted. To make the conventional format-preserving encryption system FF3-1 more secure and more difficult to break, two additional techniques are proposed as shown in Figure 10.5.

The characteristics of the suggested algorithm are as follows.

- Encrypt multiple fields with a single key
- Encrypt multiple columns in a single encoding
- The suggested approach is more secure than FF3-1 because it involves various columns.
- Non-deterministic approach, the same plaintext never yields the same ciphertext

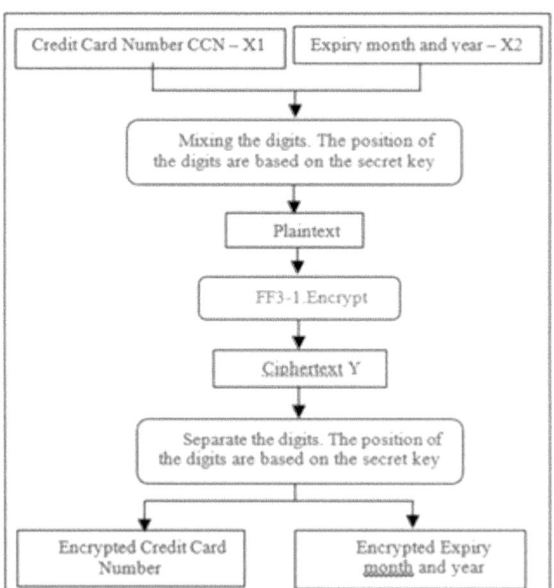

Figure 10.5 Multiple fields Format Preserving Encryption.

3.2 Python Procedures for FPE

The digits of two columns are fused, and the positions of the fusion are derived from the secret key using the Python routine fuse_digits. To conduct arithmetic operations, the keyset's hexadecimal digits are translated to decimal digits. Every digit undergoes multiplication and modulus operations to prevent the uniform distribution of locations.

```
def fuse_digits (F1, F2, k):
    # F1 – 16-digit credit card number
    # F2 – 6-digit expiry month and year of the card
    # k – 32-digit hexadecimal secret key

    # Convert F1 and F2 to lists of digits
    F1digits = list(map(int, str(F1)))
    F2digits = list(map(int, str(F2)))

    F1len = len(F1digits)
    F2len = len(F2digits )
    PTlen = F1len + F2len

    # Initialize a set to hold unique positions
    keyset = set(int(digit, 16) for digit in k)

    # Create a list of positions based on keyset values
    positions = [(element * max(keyset) % PTlen) for
                 element in keyset]
    positions = positions[:F2len]

    # Initialize the final list with zeros
    X = [0] * PTlen

    ind = 0
    for j in F2digits :
      X[(positions[ind])]=j
      ind = ind + 1
    j=0
    for ind in range(0,PTlen):
     if index not in positions:
      X[ind]=F1digits[j]
      j=j+1
    return X
```

F1 is the 16-digit credit card number, while F2 is the six-digit expiration month and year. Using the key as a base, fuse each digit in F2 to F1. The secret key determines the fusion place. To avoid duplicate digits, the hexadecimal secret key is converted to decimal numbers and saved in the Python set. The resulting list is kept on X, and the FF3-1 algorithm uses it as an input as shown in Figure 10.6.

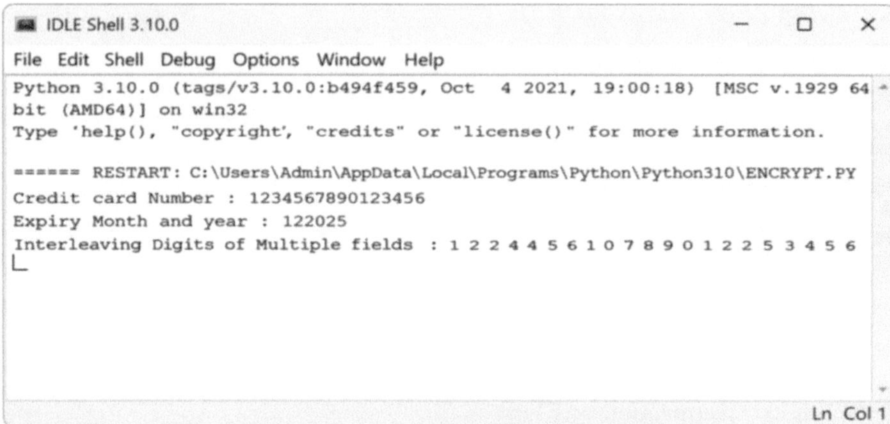

Figure 10.6 Result from the Procedure fuse_digits().

The Secret Key K, the tweak input, and plaintext X are used as input for the FF3-1 algorithm.

```
from ff3 import FF3Cipher
k = “2FE79D232DF5585D68CE47882AE256D6”
twk = “FBD09280979564”
ct = FF3Cipher(k, twk)
plaintext = “1222345610789012253456”
ciphertext = ct.encrypt(plaintext)
decrypted = ct.decrypt(ciphertext)
print(f”{plaintext} -> {ciphertext}”)
```

The output of the above algorithm is the ciphertext as shown in Figure 10.7, which is format-preserved.

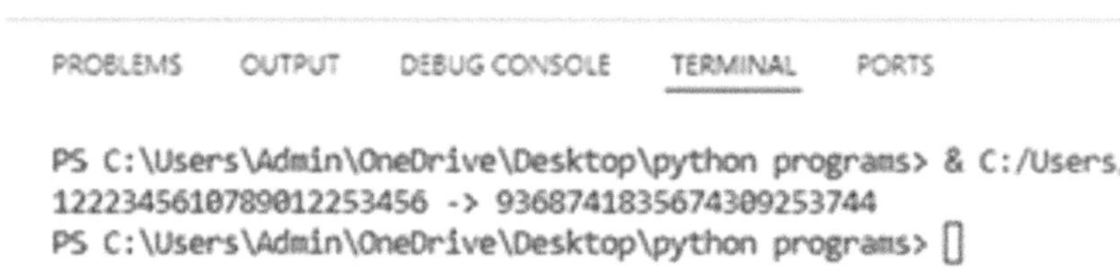

Figure 10.7 Cipertext from FF3.

The encrypted result is split into two fields using the divide_digits Python function. The output of the Format-Preserving Encryption of the FF3-1 scheme is the ciphertexts Y, whose format and data type are identical to the original fields. Number of credit cards, month, and year of expiration.

```
def divide_digits (Y, k):
    # Y is the ciphertext
    # k is the secret key

    # Initialize a set to hold unique positions
    keyset = set(int(digit, 16) for digit in k)

    # Create a list of positions based on keyset values
    PTlen = len(str(Y)) # Assume PTlen is the length of Y
    F2len = len(keyset)
    positions = [(element * max(keyset) % PTlen) for
                element in keyset]
    positions = positions[:6]

    # Initialize the final list with zeros
    X = [0] * PTlen
    F1 = []
    F2 = []
    Ydigits = list(map(int, str(Y)))

    for ind in range(0, len(Ydigits)):
     if ind in positions:
      F2.append(int(Ydigits[j]))
     else:
      F1.append(int(Ydigits[j]))

     return F1, F2

    # Example usage
    Y = 9368741835674309253744
    k = "2DE79D232DF5585D68CE47882AE256D"
    print(divide_digits(Y, k))
```

The output of the FF3-1 algorithm is an encrypted value Y that combines the encrypted segments of the credit card number, expiry month, and year. This Y is divided into Y1 and Y2 according to positions determined by the secret key as shown in Figure 10.8.

The encrypted value Y is divided into two parts: encrypted month and year of expiration (F2) and encrypted credit card number (F1). The process is the opposite of fuse_digits(). The ciphertext Y is divided into two fields, and the positions are derived from the secret key [9].

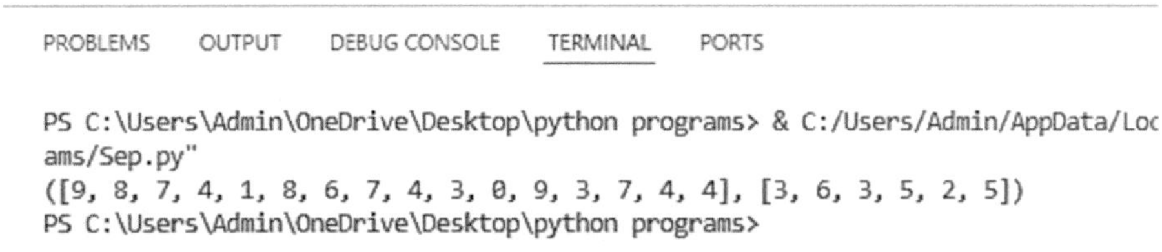

Figure 10.8 Result from the Procedure divide_digits().

The credit card number, expiration month, and year are examples of encrypted values that have the same format and data type as plaintext. Additionally, the suggested approach may be used to encrypt Aadhaar numbers, Social Security numbers, and other small domains [10].

Every program encrypts many fields with a single encryption and guarantees that the resulting text format is identical to the plaintext. To provide FPE with greater security, the suggested algorithm's complexity is raised. The suggested algorithm's security is enhanced by augmenting the quantity of encrypted fields [11, 12, 13].

Decryption is just the inverse of encryption, in which the encrypted credit card number and expiry month and year are interleaved and decrypted. After decryption, the divide_digits() procedure is used to get the original plaintext of the credit card number and expiry month and year.

The Payment Card Industry, which establishes guidelines and standards such as the Payment Card Industry Data Security Standard, guarantees the security of Cardholder Data (CHD) [14]. It is essential that upon authentication, the credit card number, expiration date, and cardholder's name be encrypted and saved. It is never possible to retain sensitive authentication data, such as PINs and CVVs, after authentication [15]. In order to test the suggested algorithm, credit card numbers and expiration dates are used. To further enhance security in subsequent work, the card number, expiration date, and cardholder's name can all be encrypted [16].

4. Conclusion

Data is encrypted using FPE while preserving its original format intact, including its length, character set, and structure. Because the encryption procedure must adhere to the format restrictions while maintaining security, this requirement adds another layer of complexity [17, 18]. High levels of security, such as confidentiality, integrity, and resilience to known-plaintext, chosen-plaintext, and brute-force assaults, are requirements for FPE. Completing these security objectives calls for increasingly sophisticated cryptographic methods and algorithms. Format-preserving encryption might be more challenging to develop and implement than traditional encryption algorithms due to these extra constraints and complexities [19, 20, 21]. The suggested method uses the FF3-1 encryption technique, providing an additional degree of security and customisation.

References

[1] Morris Dworkin. (2016). Recommendation for Block Cipher Modes of Operation: Methods for Format-Preserving Encryption. NIST Special Publication 800: 38G.

[2] Durak, F.B. and Vaudenay, S. (2017) Breaking the FF3 Format-Preserving Encryption Standard Over Small Domains, CRYPTO 2017.

[3] Orr Dunkelman, Abhishek Kumar, Eran Lambooij and Somitra Kumar Sanadhya. (2020). Cryptanalysis of Feistel-Based Format-Preserving Encryption. International Association for Cryptologic Research.

[4] Dr. Vidhya, S. (2021). Cloud security by format preserving encryption, Journal of Multidimensional Research & Review.

[5] Dworkin, M. (2019). Recommendation for Block Cipher Modes of Operation: Methods for Format-Preserving Encryption. NIST Special Publication SP 800-38G Rev. 1.

[6] Betül Durak, F., Henning Horst, Michael Horst and Serge Vaudenay. (2021). FAST: Secure and high performance format-preserving encryption and tokenization. Advances in Cryptology – ASIACRYPT (pp. 465–489).

[7] Ke Luo and Wei Gao. (2021). A study of Format Preserving Encryption algorithm Incorporating Robust Chinese Reminder Theorem. 3rd International Conference on Artificial Intelligence and Advanced Manufacture, October 2021.

[8] Yushu Zhang, Wentao Zhou, Ruoyu Zhao, Xinpeng Zhang and Xiaochun Cao. (2022). F-TPE: Flexible Thumbnail-Preserving Encryption Based on Multi-Pixel Sum-Preserving Encryption, IEEE Transactions on Multimedia, 19 August 2022.

[9] Sarah Miracle and Scott Yilek. (2022). New Algorithms and Analyses for Sum-Preserving Encryption, Advances in Cryptology – ASIACRYPT 2022.

[10] Hyunji Kim, Hyunjun Kim and Siwoo Eum. (2022). Parallel Implementation of PIPO and Its Application for Format Preserving Encryption, IEEE Access, 21 September 2022.

[11] Vidhi Bansal and Surabhi Garg. (2022). A cancelable biometric identification scheme based on bloom filter and format-preserving encryption, Journal of King Saud University – Computer and Information Sciences, September 2022, Pages 5810–582.

[12] Uddipana Dowerah, Subhranil Dutta and Aikaterini Mitrokotsa. (2023). Unbounded predicate inner product functional encryption from pairings. Journal of Cryptology.

[13] Ben Morris, Hans Oberschelp and Hamilton Samraj Santhakumar. (2023). Format preserving encryption in the bounded retrieval model. Cryptography and Security.

[14] Anselme Tueno, Florian and Kerschbaum. (2020). Efficient secure computation of order-preserving encryption ASIA CCS '20: Proceedings of the 15th ACM Asia Conference on Computer and Communications Security, October 2020.

[15] Shujie Cui, Xiangfu Song, Muhammad Rizwan Asghar, Steven D. Galbraith and Giovanni Russello. (2021). Privacy-preserving dynamic symmetric searchable encryption with controllable leakage. ACM Transactions on Privacy and Security, September, 2021.

[16] Xinle Cao, Jian Liu, Yongsheng Shen, Xiaohua Ye and Kui Ren. (2023). Frequency-revealing attacks against frequency-hiding order-preserving encryption, Proceedings of the VLDB Endowment, July 2023.

[17] Jaehyuk Lee, Sun-Young Lee, Kangbin Yim and Kyungroul Lee. (2023). Neutralization method of ransomware detection technology using format preserving encryption. Sensors.

[18] Yinghui Zhang, Robert H. Deng, Shengmin Xu, Jianfei Sun, Qi Li and Dong Zheng. (2020). Attribute-based encryption for cloud computing access control: a survey. ACM Computing Surveys, August 2020.

[19] Roman Langrehr and Jiaxin Pan. (2020). Tightly secure hierarchical identity-based encryption. Journal of Cryptology.

[20] Gilad Asharov, Hubert Chan, T.-H., Kartik Nayak and Rafael Pass. (2022). Locality-preserving oblivious RAM. Journal of Cryptology.

[21] Dr. Maheshwari, D., Hemapriyai, K.E. and Dr. Savithri, M. (2023). Design and simulation of efficient dynamic data forwarding protocol towards route optimization in VANET infrastructure. International Journal of Membrane Science and Technology 10(5): 620–626.

Chapter 11

Enhanced CBERP Detection of Attack Using AI

Savithri M.[1,*] and *Gokilamani R.*[2]

1. Introduction

The fast-expanding range of capabilities and variety of uses for mobile computing devices make MANETs so appealing to researchers and commercial companies. Alternatively, MANET nodes may not share authority but instead function as autonomous entities managing their individual assets, such as power sources and available bandwidth for data transmission [1]. One major factor preventing MANET nodes from forwarding messages to other nodes is their limited transmission bandwidth and battery life. This non-cooperative behaviour exacerbates the network's overall performance. Giving nodes an incentive mechanism to act cooperatively and forward packets over the network is essential to overcoming the lack of cooperation among nodes. These techniques can generally be divided into three categories, as shown in Figure 11.1: reputation-based, credit-based, and game-theoretic procedures. The foundation of reputation-based methods [2] and [3] is the observation of nodes' actions from a cooperative standpoint and the isolation of misbehaving nodes. There are also serious critiques of this mechanism [4]. The primary reason is that a formal evaluation of these mechanisms is not possible. The second problem is in the structure of a group of nodes plotting to maximise their utility. Their reliance on the broadcasting character of wireless networks represents the third weakness in this method. Artificial intelligence plays a vital role in various fields, including the detection of cyberattacks in wireless networks, where machine learning and deep learning are combined to reduce feature dimensions and improve

[1] Assistant Professor, Department of Data Science, CHRIST University, Bengaluru, India.

[2] Associate Professor, Department of Mathematics, Sri Ramakrishna College of Arts & Science for Women, Coimbatore, Tamil Nadu, India.

* Corresponding author: savithri.m@christuniversity.in

DOI: 10.1201/9781003635178-15

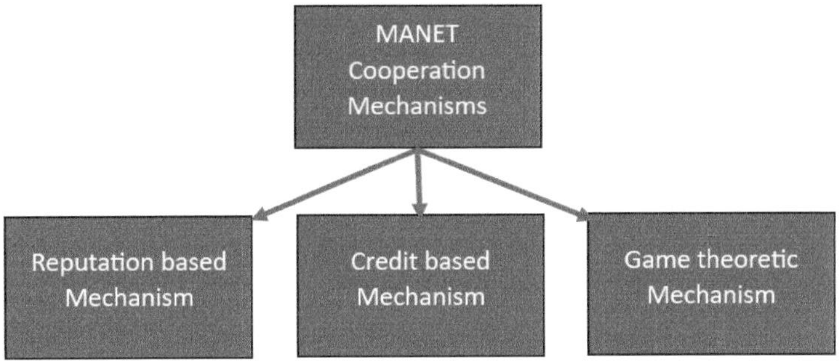

Figure 11.1 Taxonomy of MANET cooperation mechanism.

intrusion detection performance [5]. The features in the attack are identified using the K-means clustering and feature ranking.

Credits are used in the subsequent class of cooperative mechanisms as micropayments [6]. In exchange for forwarding network messages, each node is paid. Other nodes that assist in passing their messages are also paid. As a result, nodes can use credits they have previously earned from other nodes to forward messages. Each of the aforementioned outflows is a micropayment. The self-styled credit-based methods outmanoeuvre the reputation-based mechanism to get the node to collaborate. The risk of nodes acting dishonestly exists for credit-based processes [7]. Earlier attempts to stop this kind of behaviour included systems that [8] suggested guaranteeing payment security through tamper-proof hardware. The following works attempted to discard the hardware, considering appropriate restraint [9, 10]. Recent studies on cooperation in MANET have focused a lot of attention on Sprite [11], which offers an adequate level of payment security. Meanwhile, a Sprite system's mobile nodes must consume significant processing power to safeguard the incentive apparatus.

The source node obligation achieves one digital signature peer group on each of its diffused packets, and each in-between node in the packet's path that considers participating in the transmission of the source's packet must accomplish one digital signature substantiation on each packet's satisfied.

The game-theoretic mechanism, which falls under the third category of cooperation mechanisms, seeks to prevent retaliatory scenarios in the event that a node is erroneously viewed as self-interested in order to facilitate the swift restoration of cooperation. The fact that all of them have assumed flawless observation and that the majority have not considered the impact of noise on the approach design is one of the main disadvantages of this MANET method. Nevertheless, in this network, a packet may still be dropped due to transmission problems or link failures, even if a node has chosen to forward it to another node. Furthermore, it is either impossible or prohibitively expensive to achieve perfect public observation in a MANET because central monitoring is generally unavailable. Therefore, these factors render game-theoretic mechanisms inadequate for cooperative learning.

The respite of this essay is structured as follows: the analysis of the literature that bolsters the node cooperation mechanism in a MANET is covered in Section 2. In Section 3, the Scheme Description and Methodology are discussed.

The CBERP protocol design is described in Section 4. Section 5 discusses the analytical investigation of CBERP. Section 6 presents the results and scenarios from the simulation. Section 7 addresses the future scope of the effort, while Section 8 provides some concluding remarks.

2. Analysis of Literature

Most recent research efforts have been directed towards securing routing protocols with an emphasis on MANET security. The two main security concerns in MANETs are network node cooperation and the authentication procedure. To put it simply, authentication is the process by which two parties verify one other's identities. An unauthorized node may "come in" and use the network's resources with ease if authentication is lacking. If the unapproved node is a malicious node, the issue grows worse. Consequently, a system to prevent an "outsider" from linking to the network is mandatory. An authentication system that is resistant to attacks, called the Enhanced Energy Efficient Secure Enhanced Routing Protocol, was proposed in a prior study [12]. To provide a safe data transmission method against protocol attacks, a more thorough study of the ESERP protocol in a MANET has been carried out in this work than on any previous protocol. ESERP has the solution to the authentication issue. One of the protocol's shortcomings is that it does not examine a node's cooperative behaviour within the network. For this reason, the cooperation issue in the MANET is the focus of this work. The following is a summary of relevant research on MANET safety, with an emphasis on credit-based defences against DDoS attacks:

An incentive-based system, sometimes referred to as a credit-based system, provides credits to nodes as compensation for forwarding. A node does not broadcast the packets it produces without acknowledgment. In order to encourage forwarding, [13] proposed using so-called neglects, a virtual currency that functions as a per-hop payment in every packet in a secure module in every node. In order to guarantee that the precise quantity of neglects is taken out or placed, a secure module is necessary. The Packet Purse Model (PPM) and the Packet Trade Model (PTM) are the two models used to pay for packet promoting. PPM allows the individual who sent it to load the packet with many neglects by paying. When forwarding a packet, each intermediary node accepts one packet. The packet is dropped at an intermediary node if there are no more neglects there. Neglects are lost if they are still in the packet after it gets to its destination. The disadvantage of PPM is that it is challenging to calculate how many nuggets are needed to get to a particular location. The packet will be deleted, and the source will lose its investment if they underestimate this quantity. When a data packet reaches its destination, any extra data may be kept there or, if the endpoint offers information amenities, may be used to pay for those services. The issue with overestimating is that there are a number of reasons why a packet could be dropped in transit (such as buffer overflow at forwarding nodes). The destination covers the cost of the packet under the packet trade model. In order to get more neglects, each intermediary node purchases a packet from the preceding hop and vend it to the subsequent hop. The packet purse model is primarily taken into consideration since charging the destination rather than the sender can result in a network overload, and the destination receiving packets it does not want. However, this strategy may result in the loss of neglects, which a central authority would then need to reintroduce into the network.

Selfish nodes are incentivized to collaborate in the [14] Sprite incentive-based system. By uploading its receipts, a node in this system reports the communications it has received or transmitted to the Credit Clearance Service (CCS). An intermediary node gets credit for relaying a message from another node. In addition to requiring a central expert witness, Sprite also requires source routing and a public key architecture. For every transaction, Sprite uses a digital signature. The network architecture of Sprite assumes that CCS can be accessed via the Internet, which restricts the Sprite's usefulness.

A wireless health monitoring system with incentive-based route collaboration is described in [15]. To increase the reliability of message delivery, this system makes use of two cooperation protocols: the Discrete Value Cooperation Protocol (DVCP) and the Continuous Value Cooperation Protocol (CVCP). The offered and saved credit values are checked by the CVCP protocol. Approximate values like high (H), medium (M), and low (L) are used by DVCP, which is intended for smaller ad hoc networks, to represent the incentive. The priority of the message determines how credits are awarded.

Priority forwarding was introduced in [16] to encourage nodes to collaborate in forwarding network packets. Best effort and priority traffic are provided by an ad-hoc network in this design. If the source node is compensated for forwarding packets, the network grants priority forwarding by crediting intervening nodes; if not, best-effort service is rendered. This scheme's drawback is that it makes the assumption that each node in the network will stretch its best determination amenity, which is not a guarantee. The challenges of tracking other nodes' currencies within the network, the linkages between various greed measures, node behaviour, and the equilibrium of the resulting system were not addressed.

A proposed incentive structure is found in [17], wherein each node sets its own pricing for services (such as packet forwarding) that it provides to other nodes. The available bandwidth and battery power are taken into consideration while determining this price. A source node uses the cheapest transmission path. In this work, it is implicit that network nodes honestly announce matching prices.

A credit-based approach called the ad hoc-VCG method [18] addresses the collaboration issue and presents a second-best impenetrable auction type. It guarantees that a packet travels the least energy-intensive route possible. The topic of how much a node ought to charge for advancing packets raises price issues. Intermediary nodes declare genuine prices. The VCG system ensures genuine behaviour. When there is just one cheating node, the ad hoc-VCG is large. If there is cooperation among nodes to maximise payments, it could collapse. This method for mobile ad hoc network routing layer introduced a game theoretic setup where communication nodes are considered to be self-interested and have to announce their energy cost in directive to calculate a message pathway that is cost-efficient. The high overhead is another contentious issue. During the route discovery stage, it necessitates having a thorough understanding of the network topology. Lastly, it only focuses on the estimate of the payments rather than the actual delivery of the payments [19].

In the situation of delay-tolerant nets, collaboration is simulated using barter-based cooperation [20], which is not reputation- nor credit-based. The middle nodes operate a store-forward wireless mobile ad hoc network that is delay-tolerant. The nodes in this system cooperate according to a game-theoretic paradigm.

A credit-based Secure Incentive Protocol (SIP) has been suggested by Yanchao Zhang et al. [21] to simulate packet forwarding cooperation for MANETs lacking infrastructure [1]. Due to SIP's credit-based charging and rewarding system, which incentivizes nodes in an ad hoc network for the services they render or receive, even possibly self-serving but rational nodes would be encouraged to cooperate. To guard against "credit fraudulence," SIP is meticulously crafted to ensure proper charging operations and robustly withstand various threats. To lower the overhead of computation and communication, a source-controlled, session-based methodology is used. This scheme's application to unicast traffic is its restriction. SIP lacks adaptability to network-changing aspects.

A safe multilayer credit-based incentive programme was put up by Haojin et al. [22] to promote bundle forwarding activity among DTN nodes. Without depending on any tamperproof hardware, the suggested approach can be deployed in an entirely dispersed manner to frustrate different types of attacks.

Stochastic differential equations were used by Honghuing Liu et al. to provide a general mathematical framework for modelling the interplay between several underlying protocols and incentive systems. The predicted credit difference for each node in the WMN was quantified using a credit progress model, and the credit discrepancy was measured using the model of the predictable credit variation. Additionally, they suggested using distinguished pricing and demonstrated how it might lead to credit equality between nodes, making the network more resilient to varying traffic loads.

One of these credit-based mechanisms' limitations is that managing credits necessitates the use of a virtual bank. Second, a node may choose to stop cooperating and begin dropping packets if it has sufficient credits to transmit its own data. Because of the significant routing overhead, the network is unstable. Additionally, protecting messages that include credits is crucial to preventing rogue nodes from altering the value of credits. They neglected to consider the routing fairness problem, which arises when some nodes are not rewarded for certain reasons, such as location.

3. Classifications and Scope of DDoS Attack

An extensive, well-planned attack on a target device or network resource's capacity to provide services is known as a Distributed Denial of Service (DDoS) attack. An enormous quantity of hosts spread across the Internet work together in tandem to send an incredibly high volume of packets to a target system, initiating the DDoS attack. Legitimate requests will be rejected because the attack traffic uses up the target host's computer resources or the network's capacity. These assaults can have a range of effects, from slight frustration to website visitors to noteworthy financial losses for businesses that are contingent on their online presence to conduct business.

Due to the ease of access to user-friendly methods of attack that facilitate the planning and execution of large-scale DDoS attacks, DDoS attacks are projected to pose an increasingly serious threat to the Internet. With the use of these instruments, even a simple person can launch a devastating strike. Trinoo, TFN, TFN2K, Shaft, and Stacheldraht are among the existing tools that have been used in DDoS assaults on popular profitable websites like Yahoo, Amazon, and eBay. It is absurd to think that the DDoS threat can be totally eliminated without taking precautions to prevent misuse of every computer connected to the Internet. Currently, the majority of large websites address the issue by providing ample resources for crucial systems.

The days when script kiddies could transfer a programme and launch an assault on practically any website are long gone, thanks to recent developments in DDoS defence systems. Attackers now employ more intricate techniques to initiate DDoS attacks. The occurrence and size of the beleaguered networks and computers have increased intensely in DDoS attack instances despite all of the measures taken to reduce their frequency.

DoS attackers typically have a variety of motivations. Based on the attackers' motivations, we may divide DDoS attacks into five primary categories:

1) *Financial or economic gain*: Companies are quite concerned about these attacks. Attackers in this category are typically the most skilled and knowledgeable due to the nature of their motive. Financially motivated attacks are frequently the most deadly and difficult to neutralise.
2) *Vengeance*: These attackers are typically irate people, possibly with less technical expertise, who launch attacks in retaliation for perceived injustices.
3) *Ideological belief*: Attackers in this group are driven to attack their targets by their ideological convictions.
4) *Intellectual Challenge*: These attackers target the systems in order to practise and gain knowledge about how to carry out different types of attacks. Usually, they are youthful hackers who like to flaunt their skills. These days, even a computer novice may mount a successful DDoS assault by renting one of the many user-friendly attack tools and botnets available.
5) *Cyberwarfare*: These attackers are typically members of a nation's armed forces or terrorist groups, and their political goal is to target numerous vital areas of a rival nation. These strikes could target decision-making neutral departments and agencies, public and private financial institutions (such as national and profitable banks), and infrastructure related to energy and water.

Due to their scattered nature, DDoS attacks are very challenging to stop or track down. The use of spoofing, or false IP addresses by attackers to conceal their genuine identities, makes DDoS attack traceback much more challenging. Moreover, hackers can take advantage of security flaws in a lot of Internet hosts. Furthermore, there is a sharp rise in the frequency of application-layer attack occurrences. To implement a thorough DDoS defence system, one of the prerequisites is to comprehend every facet of DDoS attacks. Different DDoS attack classes have been suggested in the associated work segment. DDoS flooding attacks are categorised according to the protocol level at which they operate. As one of the most prevalent types of DDoS attacks, we primarily concentrate on DDoS protocol attacks.

As stated earlier, DDoS flooding attacks can be classified into two categories based on the protocol level they aim to attack:

Attacks using DDoS flooding at the network or transport level: The most commonly utilised protocols for these types of attacks are DNS packets, TCP, UDP, and ICMP. The following four categories of attacks fall under this category:

A.1 *Flooding Attacks*: These attacks, which include VoIP, ICMP, DNS, and spoofing/non-spoofing UDP floods, aim to block authorised users from connecting by consuming all of the target network's capacity.

A.2 *Exploitation of flooding attacks*: These attacks use explicit structures or execution flaws in some of the target's protocols to devour excessive amounts of the target's resources. Examples of these attacks are TCP SYN flood, TCP SYN-ACK flood, ACK & PUSH ACK flood, RST/FIN flood, and others.

A.3 *Flooding assaults created on reflection*: Attackers frequently send fictitious requests (such as ICMP echo requests) to the reflectors instead of sending real queries; as a result, the indicators respond to the target, using up their resources (see, for instance, Smurf and Fraggle assaults).

A.4 *Attacks using flooding and amplification*: Attackers use facilities to generate huge or numerous messages for each communication they obtain in order to maximise the amount of traffic flow that reaches the target. Botnets have remained for replication and augmentation for a very long time.

4. System Design and Methodology

An effective defence against self-serving assaults against MANET routing protocols is offered by the attack-resistant incentive system known as the Credit-Based Enhanced Routing system (CBERP). Swiftly identifying the rogue node helps the nodes delete faulty packets sooner by recognising non-cooperative activity on their part. Attacks of this kind use up the resources of the real server or of intermediary communication devices like load balancers and firewalls. Attacks on control and data traffic are handled by CBERP. Attacks are categorised according to their shared traits and objectives. For instance, the Grey Hole attack likewise drops packets, but it does so only in response to one of two factors: the sender node or time. Meanwhile, the Black Hole assault drops packets on a constant basis. From a network standpoint, when an assault begins to drop packets, it can be categorised as a black hole attack, just like the drop attack. In order for them to fall into a single group. Attacks on data traffic either cause nodes to drop packets that are going through them or cause the packets' forwarding to be delayed. The CBERP employs two situations. Attackers and pause duration are the two factors that are chosen because they are appropriate for MANETs. The first scenario aids in determining the quantity of attackers, while the second scenario assists in sending packets to their intended location in accordance with a time factor. These scenarios were chosen for the current project because the other situations would not outperform attackers and pause times.

The suggested protocol entails the following steps:

The basis and end point nodes verify that the node's classification in the node incline is correct. Positive behaviour earns more credits, while negative behaviour earns fewer credits. The source node is authenticated by the endpoint node. Each intermediary node mentioned in the packet header is verified by the endpoint node. A hop-by-hop signature offers strong defence against protocol attacks. According to the suggested protocol, before sending data to the endpoint, the sender creates a temporary key pair. The key pair consists of a public key (PS) and a secret key list (SS). By utilizing the notion of a one-way hash function, it generates the secret key list SS0 and SS1. Each entry in the list SSi needs to be hashed in order to create the public key PS. Later, when the key was generated, the sender forwarded the public key PS to the relevant addresses. The SS list is then used by the source to create the verification information. It contains the broadcast and verification data in the route request packet. Each intermediary node will use its PS to verify the source, subsequently getting the route request packet. The packet will be disposed of if the verification is unsuccessful. If not, it will be sent on. The route request will also verify the accuracy of the source verification once it arrives at its destination.

The packet is discarded if it does not work. An authentication code based on the MAC is used to ensure the integrity of the route request packet. The destination node can determine whether an intermediate node has altered any part of the route demand packet, together with the confirmation data, by comparing the MAC code with its public key. The packet is discarded if it turns out to be altered. The end point node then broadcasted the route response packet. The reply packet tracks the identical procedures as the route request packet. The suggested approach is based on the trust-based routing protocol TAODV [23]. It will find the optimal path that helps to reduce the energy drain of the node by identifying transmission of packets with the use of Artificial Neural Network (ANN), along with Support Vector Machine (SVM) based machine learning techniques.

5. Design of CBERP Protocol

Every network node maintains an additional data structure known as the Neighbour's Trust Counter Table (NTCT). Through the use of trust counters in a credit-based incentive programme, the NTCT solves the issues with secure routing. Assume that the original trust counters of the nodes {N1, N2,...} in the path from a basis S to a destination D are {TCC1, TCC2,...}

Nodes cannot be completely trusted or completely distrusted since they lack knowledge about the dependability of their neighbours at first. Route request (RREQ) packets are sent by a source S when it desires to send a packet to a destination D. A trusted credit counter (TCC) is used by each node to retain the pathway of the number of packets it has promoted along a route. With the aid of a reliable credit value counter and a credit-based incentive programme, the attacks containing the transmission tables are found. Every time node Nk accepts a packet from node Ni, node Nk increments node Ni's credit counter as

$$TCCNi = TCCNi + 1, i = 1, 2,3,4,... \quad (1)$$

Next, the values of TCCNi are added to the NTCT of node Nk. In the same manner, every node ascertains its NTCT, and eventually, the packets arrive at destination D. The end point of the RREQ communication that follows authenticates the digital signature of the intermediate nodes after examining their credit values. If the credit values of the nodes are equal to or higher than a credit threshold Ct, they are regarded as well-behaved nodes. For data transmission between nodes within a 10 m communication range, the credit threshold value Ct is initially set at 5. Afterwards, for a communication range of 20 metres and beyond, the credit threshold value Ct is raised to 10. Conversely, nodes that have credit values lower than Ct are regarded as misbehaving nodes. This lowers the control overhead because the digital signature is only verified on the offending nodes rather than on all the nodes. Nodes with trusted credit counter values lower than Ct are also blocked from receiving any more messages. Using the credit counter, the credit-based incentive mechanism (Figure 11.2) is utilised to identify malicious nodes. A new alternative path is chosen for the route discovery procedure in the event of a link failure. Because the current protocol is set up this way, a regular transmission does not result in extremely low trusted credit values. Any node identified as a black hole is removed from the route detection procedure by selecting a different path. If the foundation does not receive the RREP packet for t seconds, it will be deemed to be a route rupture or let-down. The source then inducts the route detection technique once again.

The Process of Route Discovery. Our current study uses the Energy Efficient Route Discovery technique to shift the nodes. According to the suggested protocol, a node S starts the route detection procedure by creating an RREQ packet whenever it wants to transmit a packet to an endpoint node D. It includes the source and destination IDs, a randomly generated trusted packet ID, and a MAC calculated using a common key among the sender and the end point over the trusted request ID. To ensure that the packet is authentic and hasn't been introduced into the network by a hacker, the trusted packet id is utilised. The trusted packet id confirms that only the sender or recipient is sending all of the packets that are transmitted over the designated route. When the transitional node obtains the RREQ packet for the initial time, it adds its ID to the slant of node IDs using a MAC calculated over its trusted credit counter value (TCC) and signs it using a shared key with the destination. After that, it notifies its neighbours about the RREQ. Assume that there are N1,

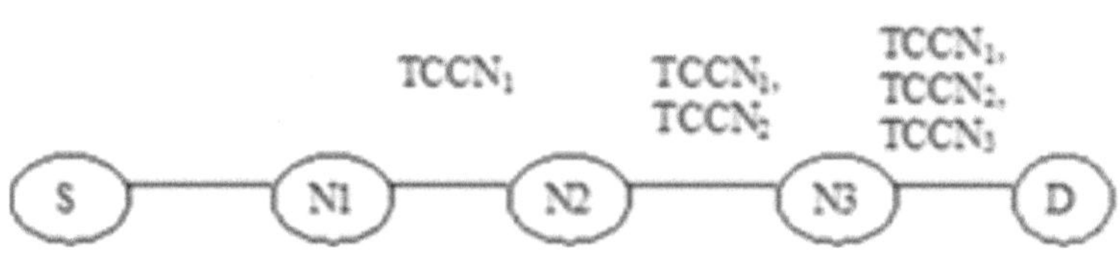

Figure 11.2 Represents the scheme of incentive.

N2,..., Nm-1 nodes separating the source S from the end point D. The procedure for requesting a route is then shown as follows:

RREQ: [S, D, MAC (TRid), TPid]

S → N1

RREQ: [S, D, MAC (TRid), TPid, MAC (TCCN1), (N1), signN1]

N1 → N2

RREQ: [S, D, MAC (TRid), TPid, MAC (TCCN1, TCCN2), (N1, N2), (signN1, signN2)]

N2 → N3

RREQ: [S, D, MAC (TRid), TPid, MAC (TCCN1,.. TCCNm-1), (N1,... Nm-1), (signN1, ...signNm-1)]

Nm-1 → D

Upon receipt of the aggregated RREQ message, the destination node recalculates the dispatcher's MAC value using its communal key, confirming first the credit values of the transitional nodes and subsequently the sender's trusted request id. Next, in order to verify that every packet was sent by the recipient initially, it checks the trusted packet ID. The receiver additionally confirms the digital signatures of every intermediate node. The threshold Ct is compared with the credit values of the intermediary nodes identify the misbehaving node. If every one of them checks out, a route reply message is generated by the destination. The destination discards the RREQ if the verifications are unsuccessful. Once more, it builds a MAC using the trusted request ID, trusted packet ID through the shared key between the dispatcher and the endpoint, and MAC computed over its trusted credit counter value (TCC) for all intermediary nodes. The trusted packet ID, the MAC of the trusted request ID, the MAC computed over TCC for all intermediate nodes, the source and destination IDs, and the cumulative route from the RREQ are all contained in the RREP and are in numerical form signed by the destination. On the return path, the RREP is sent toward the source. Because RREQ messages can only be carried up to a certain amount in the current system, there is no expansion of RREQ messages, which reduces bandwidth usage.

The Route Reply Procedure. Upon receiving the RREP packet, the midway node verifies if its ID is included in the list of IDs that the RREP has saved. It also looks up its neighbours' IDs in the list. Next, the midway node confirms that the destination node's digital signature, which is included in the RREP packet, is legitimate. The RREP packet is dropped if authentication fails. If not, the intermediary node signs it and forwards it to the following node via the opposite path.

The source first confirms that the route identified by the RREP's first id is its neighbour when it gets the RREP packet. In the event that it is accurate, the RREP packet's intermediate nodes' digital signatures are all verified. Next, it confirms that all intermediate nodes' MAC-computed TCC values are correct. Every intermediary node along the path will have its credit value adjusted or deducted based on the TCC values. The route is accepted by the source if all of these checks are successful. Along with the RREQ packet, the source also confirms the trusted request and trusted packet IDs. The absence of a replay attack is shown if the destination returned the same trusted request ID.

The route response procedure is demonstrated as below:

RREP: [S, D, MAC (TRid), TPid
MAC (TCCN1,.. TCCNm-1)
(N_1, N_2,.....N_{m-1}),(SignD)]

D $\quad$ N_{m-1}

RREP: [S, D, MAC (TRid), TPid,
MAC (TCCN1,.. TCCNm-1)
(N_1, N_2...N_{m-1}),
(SignD, signN_{m-1},...signN_2)]

N_2 $\quad$ N_1

RREP: [S, D, MAC (TRid), TPid ,
MAC (TCCN1,.. TCCNm-1)
(N_1, N_2...N_{m-1}),
(signD, signN_{m-1},...signN_2, signN_1)]

N_1 $\quad$ S

Both route request and route reply activities in this protocol require authentication. Furthermore, this cryptographic computation only needs to be performed by nodes that are stored in the current route. Thus, the suggested approach is more secure and effective.

6. Evaluation of Protocol Performance

6.1 Delay – Analytical Model

This research recommends an investigative model for computing the packet's end-to-end delay when it is transmitted over a metropolitan area network (MANET), where the MAC layer uses IEEE 802.11 DCF (Distributed Coordination Function). The delay a packet experiences in a MANET is recognized as the end-to-end suspension, and it is calculated from the moment the packet is engendered until the source node accepts an acknowledgement packet from the destination node confirming the packet was successfully received. The line-up interruptions at the source and midway nodes, the MAC delay seen at the source and intermediate nodes, and the queuing delays at the intermediate nodes make up the packet delay. A carrier-sense multiple access technique with collision avoidance and binary slotted exponential backoff makes up the distributed coordination function. The fact that 802.11 DCF can be utilised as a MAC technique for MANETs is widely recognised. In order to assess the MANETs' delay performance, the attackers in this model are changed to 5, 10, 15, 20, and 25. The network simulation parameters for the scenario are as follows: The slot time is set to 20 microseconds, and the random backoff time interval is 15 microseconds. The predictable delay met in the try state is 10 microseconds, while the expected delay in the backoff state is 15 microseconds. The Short Interframe Space (SIFS) is 10 microseconds, and the Distributed Interframe Space (DIFS), calculated as SIFS plus twice the slot time, is 50 microseconds. The radio signal transmission range is 250 meters, and the area size for the simulation is 1,000 meters by 1,000 meters. The packet size is 512 bytes, and the traffic follows a Constant Bit Rate (CBR) pattern with a transmission rate of 250 kilobits per second. The simulation will also include varying numbers of attackers, specifically 10, 20, 30, 40, and 50, to analyse their impact on the network performance.

The network in the suggested analytical model for delay consists of N nodes that are dispersed evenly and independently over a 1,000 × 1,000 m square. The transmission range of each node is assumed to be equal and is represented by r(n). The distance between nodes i and j is shown by rij. If nodes i and j can communicate directly with one another, then r ij <= r(n), designating them as neighbours. Every node has a transmission rate of Rt bits per second. Only when node i is a neighbour of node j and no other neighbour of node j is presently broadcasting with node i can node i successfully transmit a packet to node j. Every node in the network can be a packet's source, destination, or relay. Each packet is taken to have a length of L bits and a constant size. The time interval between when a packet reaches the front of the queue and when the sender receives an acknowledgement indicating that the packet was successfully received is known as the transmission delay. As a result, the delay provides the average packet service time for a node. With uniformly dispersed nodes throughout a rectangular area, this model gives an average hop count for MANETs. The average hop count is the number of hops traversed per packet between any source and any destination throughout the network. Equation (2) can be used to obtain the average end-to-end delay per packet if the average number of hops traversed per packet between a source and a destination is np.

$$D\ \text{E-TO-E} = np * E(T) \ldots \tag{2}$$

In mobile networks, the average number of hops made for each packet is provided by

$$np = E(S)/2\ r(n) \ldots\ldots \tag{3}$$

Where E(S), which is defined as the predicted hop distance within a square region of size an x b

$$E(S) = [(a3/b2) + (b3/a2) + (a/2) + (b/2)]*[\surd\ (\ a2/b2)]\ /15 \ldots\ldots \tag{4}$$

A rectangular network topology for MANETS with an area size of x b metres is considered in order to compute np, where a = 1,000 and b = 1,000 indicate its length and width, respectively. Each mobile node's radio transmission range, r(n), is taken to be 250 metres.

$$E(T) = \{(\text{Number of attackers} * (\text{DIFS} + \text{avg\_bt} + \text{DA}(i))) + (\text{SIFS} + \text{DB}(i)) + (L/Rt)\} \quad (5)$$

is the formula for the one-hop delay, where avg_bt is the random backoff time interval prior to transmission, DA(i) is the expected delay experienced in the attempt state, DB(i) is the expected delay encountered in the backoff state, and DIFS is the Distributed Inter Frame Space.

6.2 Delivery Ratio and Overhead – Analytical Model

The performance of the delivery ratio against routing disruption assaults for MANETS is studied in this paper, which describes an analytical framework for multipath routing utilising the probability of a binomial distribution. The probability that an event will occur in a single trial is denoted by p, and the binomial distribution function indicates how many times an event will occur in n independent trials. The packets are moved from the source to the destination via intermediary nodes in the suggested analytical model. To guarantee transmission success, the intermediary nodes' status is verified. The likelihood of a successful transmission is used to calculate the packet delivery ratio.

In the network scenario analysis, several parameters are defined as follows: the probability of successful data packet communication represents the likelihood that a data packet is successfully transmitted and received without errors. Computational overhead refers to the extra processing required by network nodes, typically due to protocols or security mechanisms. The node index is an identifier for a specific node within the network. The probability of failure denotes the likelihood that a data packet transmission fails. The probability of all the correctly receiving information indicates the chance that all transmitted data is received without errors. The number of equal-sized packets allocated to node i is defined by the packet allocation factor. Additionally, a represents the total number of packets transmitted for a node, while the data that reaches the destination through node i is measured separately. The probability of the event of failure to transmit the assigned packets and the probability of the event of successful attempts to transmit the packets are key metrics. The parameter p refers to the probability of getting exactly y successes in n n independent trials. The term Pr is used to determine the packet delivery ratio, and y represents the probability of success in trials out of 100. Lastly, PO denotes the overhead probability, while PNO indicates the no overhead probability.

The packets are transported from the source to the destination via intermediary nodes in the suggested analytical model for the delivery ratio. To guarantee transmission success, the intermediary nodes' status is verified. The likelihood of a successful transmission is used to calculate the packet delivery ratio. In order to increase the likelihood that data will be transmitted successfully to the intended location, the data packets are delivered utilising all independent nodes that are accessible along the path. The symbol for this likelihood is Psucc. Finding the ideal number of nodes and allocation vector given the path vector and overhead factor is the optimisation's goal. Maximising Psucc is a step in the optimisation process. The security operations, such as MAC calculation and signing, result in computing overhead (r).

Every node, denoted by the index i = (1,2,3,...., nmax), is allocated a probability of failure (Pfi) based on the likelihood that the attacks will prevent the node from transmitting. Either all information is correctly received (with probability 1-Pfi), or no information travels through the node to reach the destination (with probability Pfi). The number of equal-size packets that are assigned to node I is represented by the packet allocation vector v = [vi]. In the event that the nodes function poorly, there is absolutely no use for them. In order to maximise Psucc, the packet distribution vector v is defined as a vector of variable size n rather than fixing its size to the quantity of available nodes (i.e., nmax).

These findings lead to the following form for the allocation vector v:

$$V = (v1, v2, \ldots\ldots\ldots\ldots, vn),\ n \leq n\ \max\ldots.. \tag{6}$$

Consequently, a node's total packet transmission count is

$$a = \Sigma(i = 1)^n v_i \tag{7}$$

Given that a node with a higher failure probability cannot be given fewer packets than a node with a lower failure probability, it follows that vi ≥ vi+1 from the failure probability Pfi < Pfi+1. If Zi is the data that actually travels through nodei and viis are the data units delivered across nodei, then the likelihood of calculating the packet delivery ratio (Pr) is as follows:

$$〚Pr〛\_\ \{z\_i=0\}= p\_i\ldots. \tag{8}$$

$$〚Pr〛\_\ \{z\_i=v\_i\}= q\_i = 1-\ \ldots\ldots \tag{9}$$

Where qi is the probability of a successful attempt to transmit the packets, and Pi is the chance of an occurrence where the allocated packets are not transmitted. Here, it is assumed that the packet sent over the node is lost entirely in the event of a node failure. Thus, Psucc can be expressed as follows:

$$Psucc=f(N,a)\ \ldots\ldots\ldots\ldots\ldots\ldots \tag{10}$$

Since Pr is a discrete case, it can be approximated using a binomial distribution function with a random variable K. The number of epochs an event happens in n independent trials is specified by the binomial distribution function, where p is the chance of obtaining exactly y successes in n independent trials. The probability mass function, which provides the likelihood that a discrete random variable is precisely equal to a given value, is used to calculate Pr.

$$P(K=k) = (kn)\cdot pk\cdot(1-p)n-k\ldots\ldots \tag{11}$$

Zi~B(vi,P) is the result of the binomial distribution with parameters vi and p applied to the random variable Zi. The probability mass function can be used to determine the likelihood of receiving exactly y successes in vi tries.

Equation (12) is produced by substituting Equation (10) into Equation (11).

The formula for Psucc is

$$P(K \leq k) = \sum i{=}0k(in)\cdot pi\cdot(1-p)n-i. \quad (12)$$

In the case where

$$p = 0.8 \text{ and } D\ E\text{-}TO\text{-}E = np * E(T) \quad (13)$$

As a result,

$$PDR = 5Pr \quad (14)$$

can be used to express the packet delivery ratio (PDR) of the path.

The explanation for 5Pr in equation 14 is that, five times out of ten, a packet will be transported securely. Owing to the attackers' and nodes' mobility, the success rate is 5/10, or 50%. In our work, we used the analytical procedure ten times. PDR thus equals 5Pr.

The formula for calculating the overhead probability (PO) and no overhead probability (PNO) is as follows: PO = Pr {OH=1} PNO = Pr {OH=0} = 1 – PO, where Pr is the packet delivery ratio probability and OH is the overhead component.

6.3 Model and Parameters for Simulation

The suggested approach is replicated using the network simulator NS2. The mobile hosts' channel capacity in the simulation is fixed at 2 Mbps. For wireless LANs, the MAC layer protocol of IEEE 802.11's distributed coordination function (DCF) is utilised. It is capable of informing the network layer when a link breaks. For 50 simulation seconds, 100 mobile nodes travel within a 1,000 m × 1,000 m square area in the CBERP simulation. It has been considered because a square zone measuring 1,000 metres by 1,000 metres offers good coverage for the MANET area. A sample of 100 mobile nodes and 50 seconds was taken; this number can be expanded in the future if necessary. Every node is assumed to move independently and at the same average speed. The transmission range of 250 metres is shared by all nodes. The lowest and maximum speeds in the simulation are 5 and 10 m/s, respectively. Constant Bit Rate (CBR) traffic is being simulated. Table 11.1 represents the settings of the simulation used.

Table 11.1 CBERP.

No. of Nodes	**100**
Area Size	1000 × 1000
MAC	802.11
Radio Range	250 M
Simulation Time	50 Sec
Traffic Source	CBR
Packet Size	512
Speed	5 m/s to 10 m/s
Misbehaving nodes	5, 10, 15, 20, 25

6.4 Performance Metrics

The performance of MANET can be assessed using a wide range of performance indicators, including throughput, control overhead, end-to-end delay, delivery ratio, number of missed packets, route acquisition time, average hops per route, etc. The following performance measures are better suited for assessing the network's performance in light of the problem definition:

Control overhead: The total number of received data packets normalised by the total number of routing control packets is the definition of control overhead.

Total number of packets transmitted minus the total number of data packets sent equals overhead.

Average end-to-end delay: This is the total amount of time it takes for all data packets to survive as they travel from their sources to their destinations.

Time of data packets received at the destination minus time of data packets supplied at the source equals the end-to-end latency.

The percentage of successfully received packets to all packets delivered is known as the average packet delivery ratio.

Delivery ratio is intended as follows: (Total number of packets conveyed/Number of packets effectively received at the end point)

The failure of one or more sent packets to reach their destination is known as packet loss, and it is measured by the packet loss ratio.

Number of packets lost/Number of packets sent is the packet loss ratio.

The following section presents the simulation results for the two situations. In the occurrence of a spiteful node environment, the performance of CBERP is compared with the TAODV protocol.

7. Results and Simulation Scenario

A comparison is made between the current TAODV (Trusted AODV) protocol and the proposed CBERP protocol. The current TAODV protocol has a number of noteworthy features, including the following: (1) nodes carry out trusted routing behaviours primarily based on their relationships of trust; (2) a node engaging in malicious behaviour will eventually be identified and blocked from access to the entire network; and (3) system performance is enhanced by avoiding the creation and verification of digital signatures at each routing hop.

We change the number of attackers in our experiment to 5, 10, 15, 20, and 25.

The packet delivery ratio findings for the mischievous nodes 5, 10, .25, and 100 nodes are displayed in Figure 11.3. The convergence of the analytical and simulation results, with a difference of 10–20%. As the number of assailants rises, the delivery ratio falls.

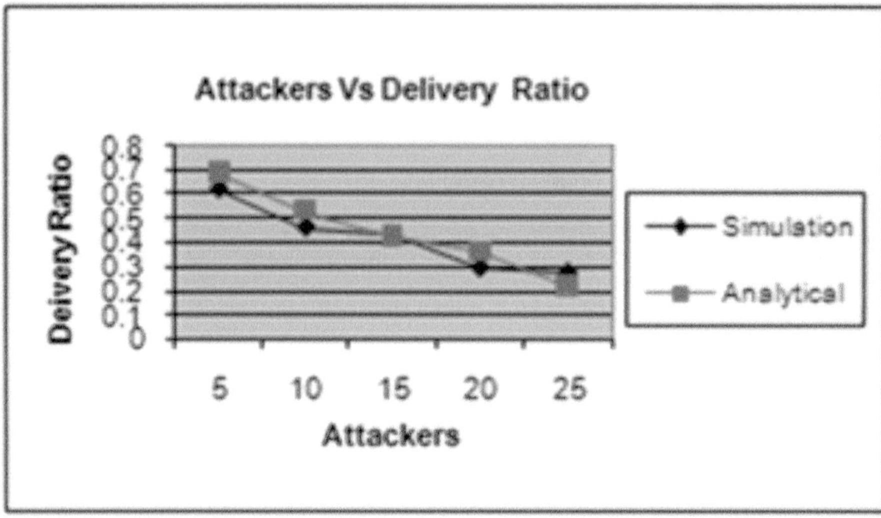

Figure 11.3. Comparison of attackers and delivery ratio.

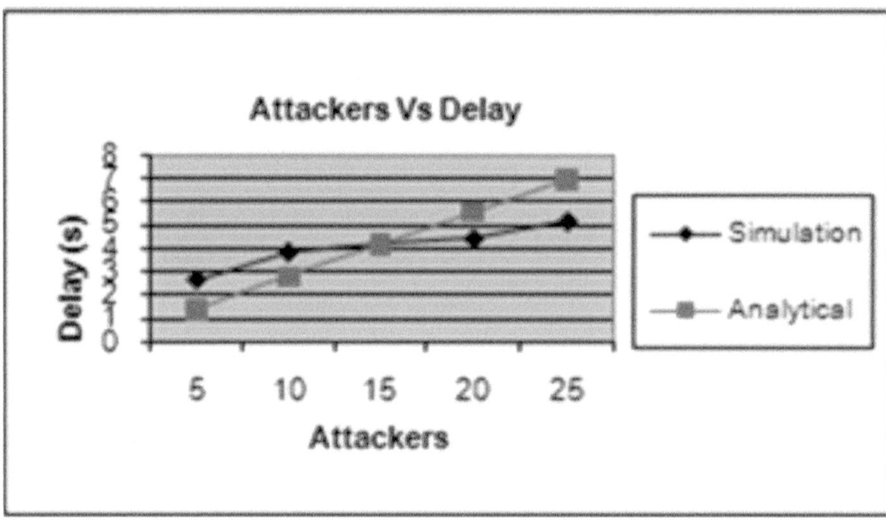

Figure 11.4. Comparison of attackers with delay.

The results of the delay for the misbehaving nodes 5, 10, .25, and so on for the 100 nodes scenario are shown in Figure 11.4. It is evident that there is a 20–30% gap between the analytical and simulation results. The more attackers there are, the longer the delay.

The chance of trial success is displayed in Figure 11.5. The accompanying figure illustrates how the probability of success rate gradually drops as the number

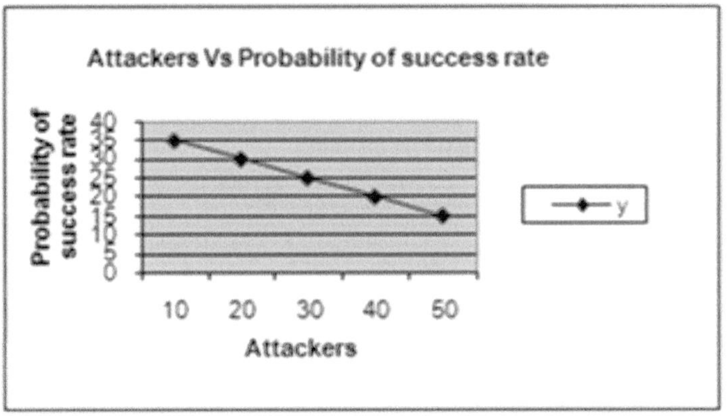

Figure 11.5 Attackers vs. probability of success rate.

Table 11.2 Comparison delivery ratio's analytical and simulation result.

No. of Attackers	Simulation Values	Analytical Values
5	0.62414	0.69324
10	0.46826	0.53727
15	0.438508	0.42707
20	0.303881	0.36632
25	0.283143	0.22132

of attackers grows. Because of the nodes' mobility, the analytical and simulation results typically disagree. Analytical results show value fluctuations in successful time intervals, while simulation values rise or fall as a result of node mobility in MANET and unpredictability. The analytical and simulation values are tabulated in Table 11.2.

A. Scenario 1: The simulation settings vary the number of misbehaving nodes to 5, 10, 15, 20,..., 50.

The average packet delivery ratio results for the misbehaving nodes 5, 10, 15, 20,..., 50 for the scenario with 100 nodes are displayed in Figure 11.6. Since CBERP offers both security and dependability qualities, it is evident that it achieves a higher delivery ratio than TAODV. By maintaining the route's link reliability, CBERP outperforms TAODV in terms of packet delivery ratio. Because of the increased node mobility in the presence of 50 attackers in particular, both the CBERP and TAODV protocols have extremely low delivery ratios.

The average end-to-end latency findings for the misbehaving nodes 5, 10, 15, 20,... 50 are displayed in Figure 11.7. Based on the data, CBERP's delay is somewhat less than TAODV's due to authentication procedures. Due to the increased node mobility in the scenario with 25 attackers, both the CBERP and TAODV protocols experience significant delays.

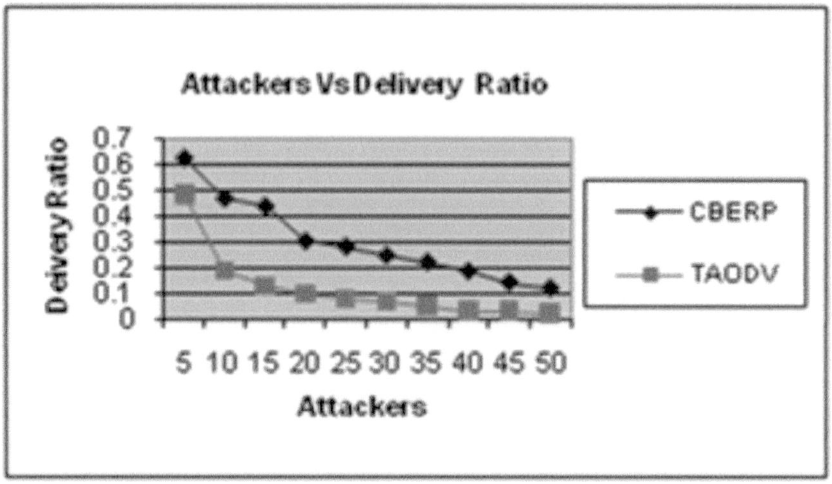

Figure 11.6 Attackers Vs Delivery Ratio.

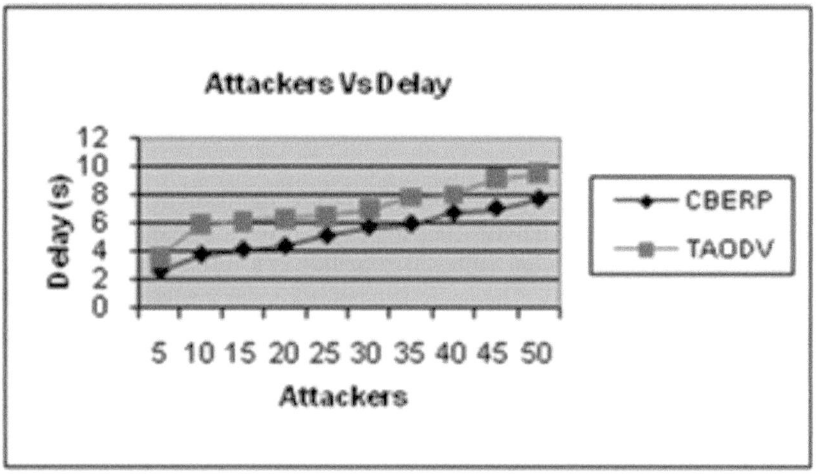

Figure 11.7 Attackers Vs Delay.

The routing overhead findings for the misbehaving nodes 5, 10, 15, 20,... 50 are displayed in Figure 11.8. Based on the findings, the CBERP protocol has lower routing overhead than the TAODV because it uses routines for route re-discovery and a credit-based incentive system.

The packet loss ratio findings for the misbehaving nodes 5, 10, 15, 20,... 50 are displayed in Figure 11.9. According to the results, because the CBERP protocol uses a credit-based incentive system, it has a lower packet loss ratio than the TAODV.

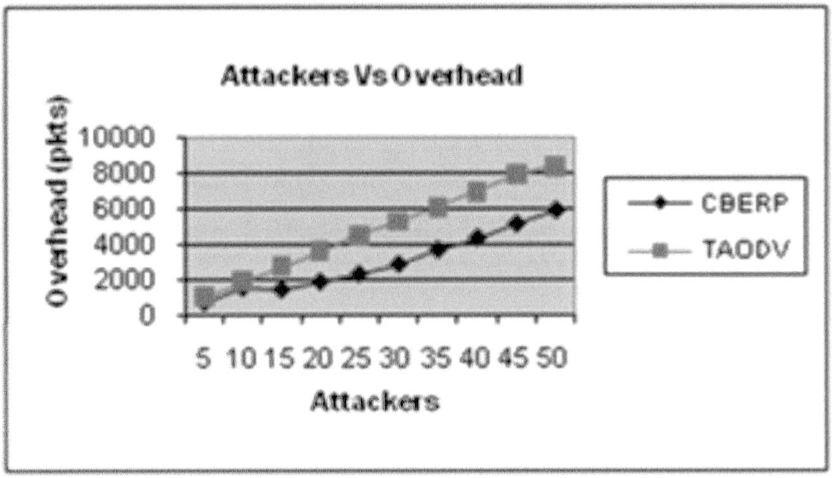

Figure 11.8 Attackers Vs Overhead.

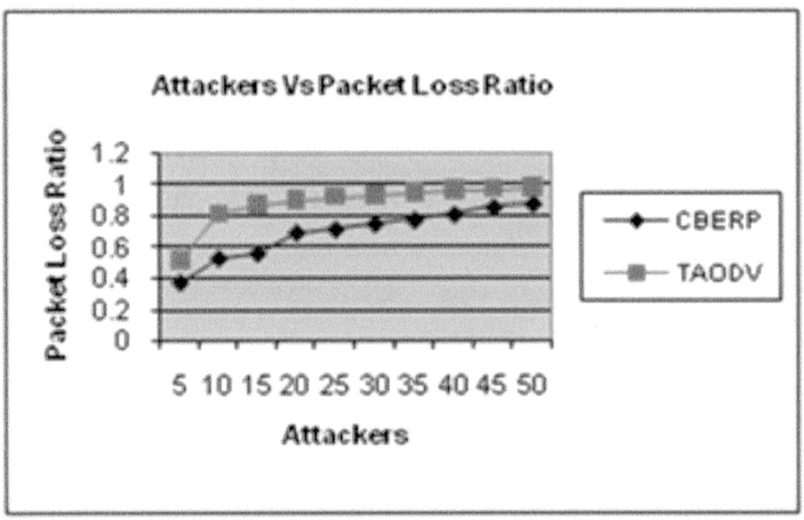

Figure 11.9 Attackers Vs Packet Loss Ratio.

B. Scenario 2: In the second scenario, there are five attackers in the simulation settings, and the pause times are changed to 10, 20, 30, 40, and 50.

The average packet delivery ratio findings for the pause times of 10, 20, and 50 for the scenario with 100 nodes are displayed in Figure 11.10. CBERP outperforms TAODV in terms of robustness and delivery ratio while having a longer pause period.

The average end-to-end delay findings for the pause times 10, 20, .50, are displayed in Figure 11.11. According to the results, CBERP exhibits superior path optimality and has a smaller delay than TAODV.

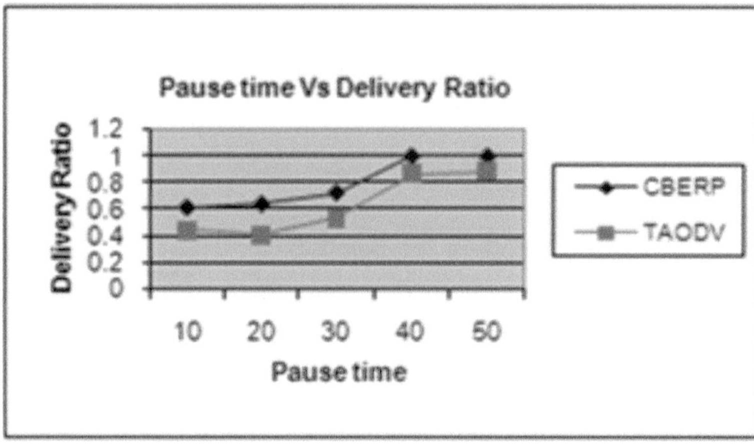

Figure 11.10 Pause time Vs Delivery Ratio.

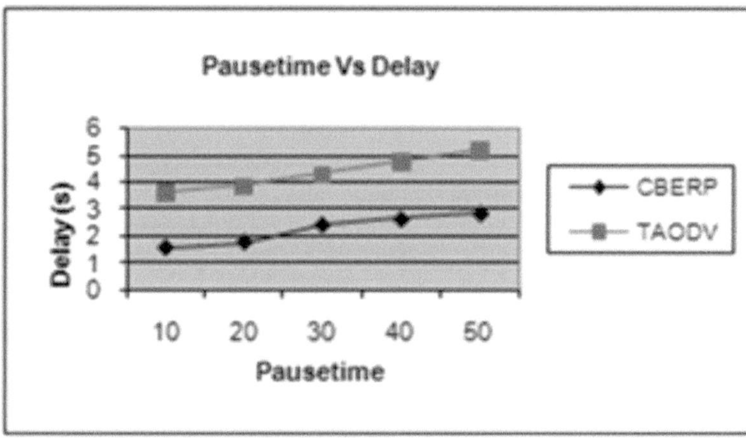

Figure 11.11 Pause time Vs Delay.

The routing overhead findings for the pause times 10, 20, .50, are displayed in Figure 11.12. According to the data, CBERP has less routing overhead than TAODV since it continues to use a credit-based incentive programme to cut down on superfluous overhead.

The packet loss ratio results for pause times of 10, 20, and .50, are displayed in Figure 11.13. According to the results, CBERP has a lower packet loss ratio than TAODV because it continues to use a credit-based incentive programme to reduce unnecessary packet loss.

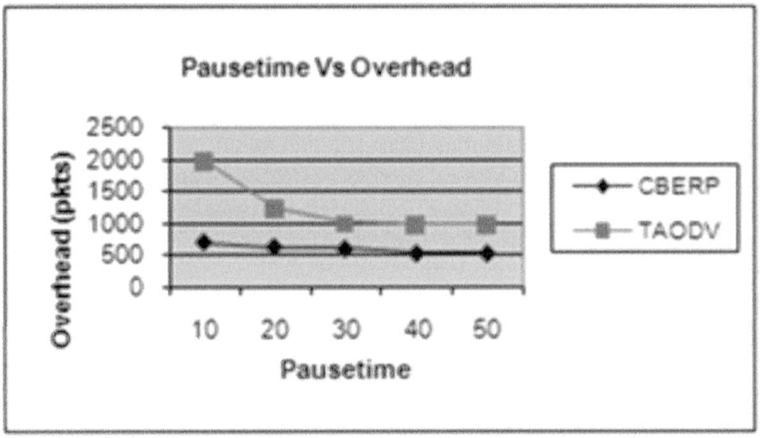

Figure 11.12 Pause time vs overhead.

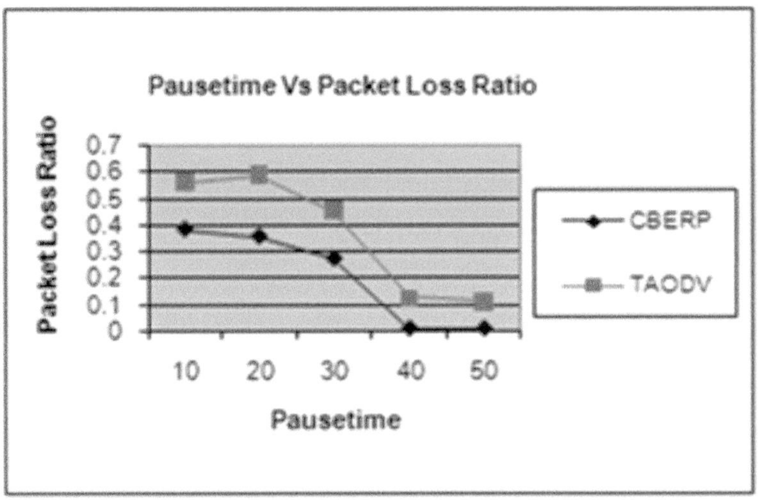

Figure 11.13 Comparison of Pausetime with PacketLoss.

8. Conclusion

The creation and assessment of the Credit-Based Enhanced Routing Protocol (CBERP), a novel mobile ad hoc network routing protocol that improves node cooperation and offers secure routing. The TAODV routing protocol is the foundation of the design. There is a comparison between game-theoretic mechanisms and reputation-based strategies. By swiftly identifying malicious nodes and confirming the digital signatures of all intermediate nodes, CBERP helps the nodes drop faulty packets early. It rewards well-behaved nodes by increasing the credit counter and penalises misbehaving nodes by decreasing the trusted credit counter. As a result, it stops hostile or compromised nodes from targeting the routes with uncompromised nodes. Based on the two scenarios, the suggested protocol CBERP performs 75–80% better overall than the current protocol TAODV, which performs 55–65% better. (1) Attackers; (2) Interval of pause. The efficient protection against data traffic and control traffic attacks in MANETs is the main goal of the proposed protocol. Mobile nodes with limited energy resources have the computational overhead quantified. CBERP demonstrates that the current protocol effectively protects against protocol attacks in MANETs while providing a secure data transmission method.

References

[1] Gupta, P. (2016). A literature survey of MANET. International Research Journal of Engineering and Technology.
[2] Mohammad Amir Khusru Akhtar, G.S. (2021). Enhancing Cooperation in MANET using neighborhood compressive sensing model. Egyptian Informatics Journal.
[3] Sandeep Kumar, S.K. (2020). Study of MANET: Characteristics, challenges, application, routing protocol and security attacks. International Journal of R&D in Engineering Science and Management.
[4] Zheng, S.X.X.H. (2012). Energy and latency analysis for in-network computation with compressive sensing in Wireless Sensor Networks. Proceedings IEEE INFOCOM, pp. 2811–2815.
[5] Mohamed H. Behiry, M.A. (2024). Cyberattack detection in wireless sensor networks using hybrid feature reduction technique with AI and Machine Learning methods. Journal of Big Data, pp. 1–39.
[6] Ammar Kamel, M.H.Z.S.B.Z.M.A. (2021). A credit-based system for traffic routing in support of vehicular networks. Journal of Case on Information Technology.
[7] Yuxin Mao, C.Z.J.Q.X.Z. (2020). A fair credit-based incentive mechanism for routing in DTM-based sensor network with nodes' selfishnesss. EURASIP Journal on Wireless Communications and Networking.
[8] Chinchu, V.S.M.J. (2014). Trust based secure payment scheme for multi-hop wireless networks. IOSR Journal of Computer Engineering.
[9] Qingfeng Jiang, C.M.Z.T. (2016). A credit-based congestion-aware incentive scheme for DTNs. Information and Communications Technology.
[10] Cheng, R.Z.N.W.H. (2014). Service-oriented node scheduling schemes with energy efficiency in wireless sensor networks. International Journal on Distribution of Sensor Networks.
[11] Yuma Shibasaki, K.L.K.S. (2022). A Communication-Efficient Secure Routing Protocol for IoT Networks. Sensors (Basel).
[12] Prasad R, R.S. (2022). Enhanced energy efficient secure routing protocol for mobile ad-hoc network. In Global Transactions Proceedings.
[13] Srdjan Capkun, J.-P.H.L.B. (2017). Mobility helps security in ad hoc networks. Journal of Mobile Adhoc Network.
[14] Wenjing Lou, W.L.Y.F. (2017). A Secure incentive protocol for mobile ad hoc networks. Wireless Networks.
[15] Varshney, U. (2008). Improving wireless health monitoring using incentive-based router cooperation. Computer, pp. 56–62.
[16] Jayson k. Jayabarathan, S.R.A.R.S. (2016). QoS enhancement in MANETs using priority aware mechanism in DSR protocol. EURASIP Journal of Wireless Communications and Networking.
[17] Younghwan Yoo, S.A.D.P.A. (2010). Impact of a simple load balancing approach and an incentive-based scheme on MANET performance. Journal of Parallel and Distributed Computing.
[18] Rama Abirami, M.S.K. (2017). Preventing the impact of selfish behavior under MANET using Neighbor Credit Value based AODV routing algorithm. Indian Academy of Sciences.
[19] Dimitra G. Kampitaki, A.A. (2023). Selfishness in mobile ad-hoc networks: a literature review on detection techniques and prevention mechanisms. IEEE Access.
[20] Minwook Kang, Y.W.C. (2020). An improved hybrid routing protocol combining MANET and DTN. Electronics.

[21] Yanchao Zhang, W.L.Y.F. (2004). SIP: A secure incentive protocol against selfishness. IEEE Communications Society.
[22] Haojin Zhu, X.L.R.L.Y.F.X.S. (2009). SMART: A secure multilayer credit-based incentive scheme for delay-tolerant networks. IEEE Transactions on Vehicular Technology.
[23] Shaik Shafi, S.S. (2023). Machine learning and trust based AODV routing protocol to mitigate flooding and blackhole attacks in MANET. Procedia Computer Science, pp. 2309–2318.

Chapter 12

A Transformer-Assisted Reversible Data Embedding Framework for IRIS Images Using STC and Hashing

Ms. S. Selvarani[1,*] and *M. Mary Shanthi Rani*[2]

1. Introduction

Data hiding has become one of the most important techniques in digital security. It has various applications such as steganography, watermarking, and covert communication. The main objectives of data hiding are: invisibility, meaning the hidden data is not perceptible in sensory data (cover-elements); capacity, which refers to embedding a large amount of hidden information while preserving the quality of the cover media; and robustness, whereby the hidden data can be reliably extracted under varying noisy and worst-case circumstances [1]. However, most conventional data hiding-based schemes change the host media irretrievably, which is not applicable for the fields of complete recovery, for example, medical imaging and biometric security. To eradicate this problem, Reversible Data Hiding (RDH), also known as Lossless Data Hiding (LDH), has become widespread. RDH methods

[1] Assistant Professor, Department of Computer Applications, Fatima College, Mary Land, Madurai, and Research Scholar, Department of Computer Science and Applications, The Gandhigram Rural Institute (Deemed to Be University), Dindigul.

[2] Professor, Department of Computer Science and Applications, The Gandhigram Rural Institute (Deemed to Be University), Dindigul.

Email: drmaryshanthi@gmail.com

* Corresponding author: rani.s.selva@gmail.com

DOI: 10.1201/9781003635178-16

tend to recover the original image and embedded data exactly, which is very crucial in applications where slight alterations are not tolerable, where the exact reconstruction of omni media is a critical aspect, such as forensics and biometrics [2, 3].

Iris as a biometric modality exhibits high uniqueness and stability that is non-intrusive and is used in real scenarios. However, it is quite difficult to hide private information in iris images, because the quality of the iris image can be directly reflected in the recognition accuracy. Conventional RDH methods, e.g., histogram shifting, difference expansion, and interpolation-based embedding, tend to adjust pixel values rather mindlessly, thus distorting valuable texture information essential for accurate iris recognition [4–6]. In this paper, to address this problem, we present a solution known as the Transformer-Based Reversible Data Hiding Scheme for Iris Images, which is integrated with Syndrome Trellis Encoding (STE) and an adaptive region-aware distortion estimation method. This is what makes our work interesting: a transformer-like attention mechanism for embedding. A location-aware cost map based on the gradient intensity and semantic edge is constructed by using the smart priority assignment for the embeddings on different regions with negligible degradation on the biometric performance. Furthermore, in order to maintain data integrity and detect tampering, the ROI is hashed by the cryptographic hash function before hiding. It enables secure authentication and error detection as the data is modified. STC can be used for LSB substitution (Least Significant Bit) and MR embedding with minimum bit flip [7].

Extensive experimental results are shown on the CASIA-IrisV2 dataset, one of the most popular benchmarks in biometric literature [8]. Performance is evaluated by PSNR, SSIM, and iris recognition rate under various hiding payloads. Experimental results demonstrate that the proposed scheme outperforms the existing histogram shifting and STC-based methods, especially in high payload, and is an effective approach for secure and reversible biometric data embedding. We organise the rest of the paper as follows: in Section 2, we review some related RDH techniques. In Section 3, we describe the methods, and in Section 4, we present our proposed Transformer-aided framework. Section 5 describes the experimental results and analysis. The last section presents some concluding remarks and possible development points for future work.

2. Literature Review

Reversible Data Hiding (RDH) has been extensively studied for embedding confidential information into host images while enabling full recovery of the original. Techniques have evolved from classic pixel-level methods to advanced learning-based strategies. This section surveys pivotal developments, with an emphasis on biometrically sensitive modalities such as iris images.

RDH approaches (Figure 12.1) are commonly grouped into pixel-based, block-based, interpolation-based, and transform-domain categories. Pixel-level techniques such as LSB substitution, prediction-error expansion, and histogram shifting offer high embedding capacity, but often compromise important structural features in biometric images, such as iris textures. The CPBS [9] embeds decisions

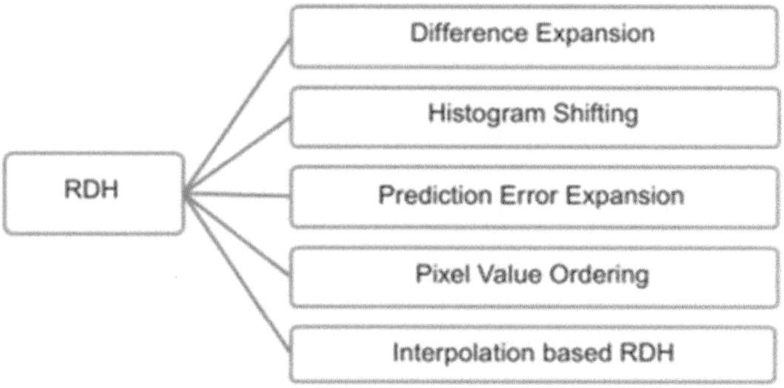

Figure 12.1 Categories of RDH.

bound with the local complexity of the image. State-of-the-art transformer-based techniques provide improved region sensitivity. Lightweight Vision Transformer (ViT) architectures (e.g., MiniViT and MobileViT) have demonstrated so far promising results in region-aware imaging tasks in the form of an alternative direction for embedding strategies that cater to semantic textures that are of most concern with respect to maintaining the viability of iris recognition [10].

Pixel differencing encodes information based on the relationship between neighbouring pixels. Tai et al. proposed histogram-shifting, which is based on pixel difference spread to enhance the payload while avoiding large distortion [11]. The interference cancel model of Li et al. was used to suppress the salt-and-pepper noise in RDH embedding [12]. Celik et al. refined these concepts via entropy coding with arithmetic difference and prediction models [13, 14]. In block-based methods, the images are divided into local processing blocks. Hong et al. introduced HSPE (Histogram Shifting of Prediction Errors) for shifting the histograms of prediction errors [15]. Ni et al.'s zero/peak histograms are now a cornerstone technique of RDH [16]. Zeng et al. found enhanced flexibility with adaptive scanning and tile-based methods [17], and Fallahpour et al. concentrated on high-frequency embedding areas for capacity optimization [18].

For domains demanding high fidelity (e.g., medical imaging), transform-domain RDH is preferred. Zhang and Hong et al. utilized DCT and wavelet domains to maintain structural integrity [19]. Huang et al. adapted gradient-aware histogram shifting for high-bit-depth medical images [20], and Chen et al. employed JND models to fine-tune distortion thresholds [21]. Khan et al. introduced RW-HPBS for authentication-targeted embedding using wavelet-based block selection [22]. While histogram methods generally maintain low distortion, they often lack capacity. Biometric-specific solutions address this gap. Whitelam et al. applied multi-layer watermarking to protect iris and facial data simultaneously [23]. Some works restrict embedding to specific channels (e.g., blue iris bands) to reduce perceptual impact but often neglect recognition metrics.

RDH tailored for iris images has been less explored but is promising. Yalman et al. proposed RCM with Neighbour Mean Interpolation (NMI) to achieve high-capacity reversible embedding without compromising biometric details [24]. An enhanced version with region-specific weighting improved recovery fidelity [25]. Recently, RDH has been widespread with deep learning and the transformer architectures. A 2023 paper reviewed deep-learning-based RDH for iris images, which included a distortion function, SVR, and histogram shifting [26]. Another interesting technique is BioDeepHash, which employs deep and hashing functions, cryptographic chopping for secure biometric template protection, providing substantial FAR reduction on iris datasets [27]. In a preprint of 2025, the privacy-preserving iris recognition based on neural style transfer was investigated, which worked under the assumption that identifiability-styled features should be obfuscated while preserving the performance of recognition [28]. This is not a specific RDH, but it demonstrates exceptional embedding strategies that would be pertinent to iris security. This recent research results indicate that there is an emerging trend of these secure, reversible, region-sensitive, and learning-embedded RDH systems as exactly what this work has pursued, specifically by exploiting Transformer-based attention, STC embedding, and secure hashing for better iris data embedding.

3. Materials and Methods

The proposed Transformer Abetted Reversible Data Embedding Framework is designed for protected and indiscernible embedding of sensitive data within iris biometric images. The experimental setup utilizes the CASIA-IrisV4 dataset (Figure 12.2), which offers high-quality near-infrared iris images under various lighting and acquisition conditions. Initially, iris images undergo standard pre-processing, including segmentation, normalization, and enhancement to isolate the iris region and suppress noise.

Feature-rich patches are then extracted and passed through a Vision Transformer (ViT)-based embedding attention module, which identifies optimal embedding regions guided by attention weights. This region-aware strategy helps reduce embedding distortion. Subsequently, Syndrome-Trellis Codes (STC) are employed

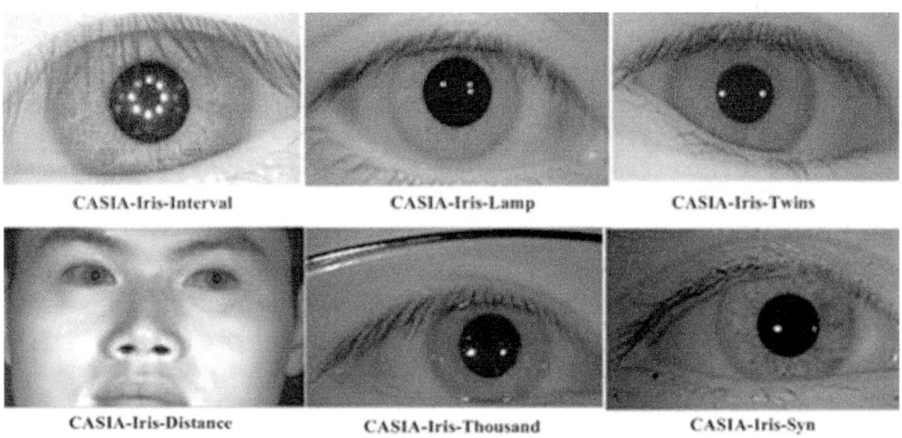

Figure 12.2 Sample IRIS Images from CASIA-Iris V4.

to encode the payload efficiently while minimizing modification rates. To enhance data integrity and prevent tampering, the embedded data is hashed using a SHA-3 512-bit hashing function before insertion. All experiments are implemented using colab, and evaluations are conducted in terms of PSNR, SSIM, Bit Error Rate (BER), and biometric recognition rate to evaluate trustworthiness, reversibility, and high security.

4. Proposed Methodology

This paper introduces a novel, progressive, RDH scheme tuned for biometric iris images. The main challenge is to embed rich supplemental data into the host stimulus and preserve the trust and recognition performance of the iris-based authentication system. The proposed method integrates STC with a transformer-based cost prediction and a secure hashing approach for integrity verification.

4.1 Dataset and Experimental Setup

As a validation, experimental results were tested on methods in the CASIA-IrisV2 public iris image database, a well-known iris recognition standard. ASIST2014 dataset: From this whole dataset, a total of 600 irises in grayscale, 60 of the same persons, 10 per person. We captured images using near-infrared light and of the same 640 480-pixel bitmap format to ensure a fair comparison in terms of reversibility, perceptual quality, and biometric performance of the embedding scheme.

4.2 System Architecture Overview

The proposed framework is composed of four sequential modules:

- *Edge-Aware Gradient Pre-processing*: Iris images are first pre-processed using Sobel and Laplacian filters to generate gradient maps. These maps highlight flat or smooth regions with lower feature concentration, which are more suitable for embedding without affecting iris recognition reliability.
- *Transformer-Based Cost Map Estimation*: A lightweight Vision Transformer (MiniViT) is employed to model both local and global dependencies in the texture. The transformer's attention mechanism is trained to assign lower embedding costs to regions with low texture complexity while penalizing feature-rich zones. The resulting adaptive distortion cost map guides the embedding process by identifying optimal, low-risk regions.
- *STC-Based Data Embedding*: STC-Embedded Data based on identified zones, payload is embedded using Syndrome-Trellis Coding (STC). This encoding model keeps the number of replacing pixels to add the data as low as possible and, as a result, keeps the host image high fidelity. Here, the embedding is dynamically adjusted to the transformer-produced cost map in order to minimize the visual and biometric distortion.

- *SHA-256-Based Data Integrity and Authentication*: Authentication and data integrity based on SHA-256: To help guarantee the integrity and authenticity of data, the embedded payload is protected by a SHA-256 hash. ROI hash is calculated and saved together with the payload. During retrieval, it recalculates and checks this hash to identify any hash tampering or loss of data.

4.3 Data Extraction and Reversibility

At the receiver end, the embedded image undergoes reverse processing. The STC decoder recovers the hidden payload with minimal bit errors. The ROI hash is verified against the embedded hash to confirm data integrity. Additionally, the system ensures full reversibility; the original image is restored exactly, with LSB (Least Significant Bit) restoration ensuring no degradation of the biometric template.

Steps:

1. Input Iris Image (CASIA – Iris V2 Data Set)
2. Pre-processing - Resize, Normalize, Localization
3. Region Selection: Distortion Function
4. STC Encoding for Data Embedding
5. Reconstruct Image with Embedded Data
6. Data Extraction and Image Recovery

5. Performance Evaluation

The proposed Transformer-assisted RDH system demonstrated strong performance across embedding capacity, visual quality, and recognition accuracy. PSNR consistently exceeded 35 dB, and SSIM remained above 0.96, indicating high perceptual and structural fidelity. Iris recognition accuracy stayed above 95% even at a payload of 0.2 bpp, confirming minimal impact on biometric utility. The system achieves a favorable balance between capacity and distortion while ensuring reversibility. Additionally, SHA-256 hashing enhances data authenticity and tamper detection, making it suitable for secure biometric applications.

6. Results and Discussion

As payload increases, recognition rates decrease, as expected due to distortion from embedding. The Proposed Method shows minimal accuracy drop, performing better than traditional methods like HILL, HUGO, and EMD. The Original (no embedding) line stays flat at 99.2%, acting as a benchmark.

PSNR vs Payload: This chart evaluates steganographic performance by showing how PSNR decreases as embedding payload increases on the CASIA-IrisV2 dataset. PSNR reflects image quality; higher values indicate less distortion. All methods show declining PSNR with increased payload, indicating quality loss. The Proposed Method consistently maintains the highest PSNR, showing better image preservation, whereas EMD performs the worst with significant degradation (Figure 12.3).

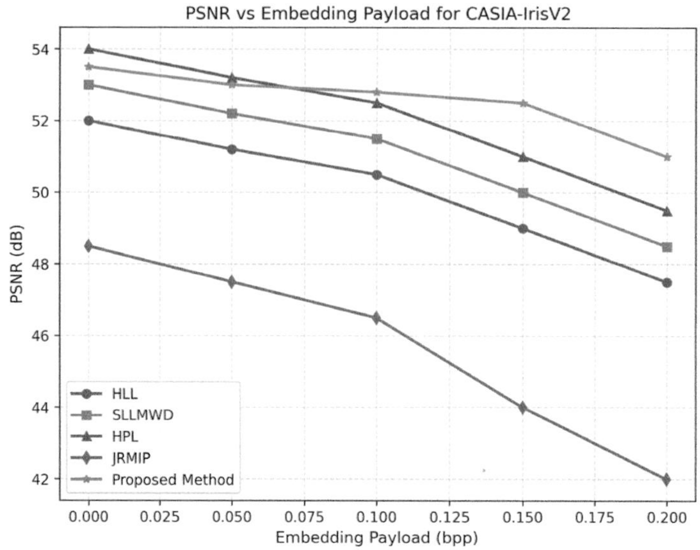

Figure 12.3 PSNR vs Payload.

SSIM vs. Payload: SSIM (Structural Similarity Index) measures perceived image quality, with values closer to 1 indicating better similarity. The graph shows SSIM versus embedding payload for various steganographic methods on the CASIA-IrisV2 dataset. As payload increases, SSIM declines for all methods due to embedding distortion. The Proposed Method consistently achieves the highest SSIM, preserving structural integrity, while EMD shows the steepest drop, indicating more visual degradation (Figure 12.4).

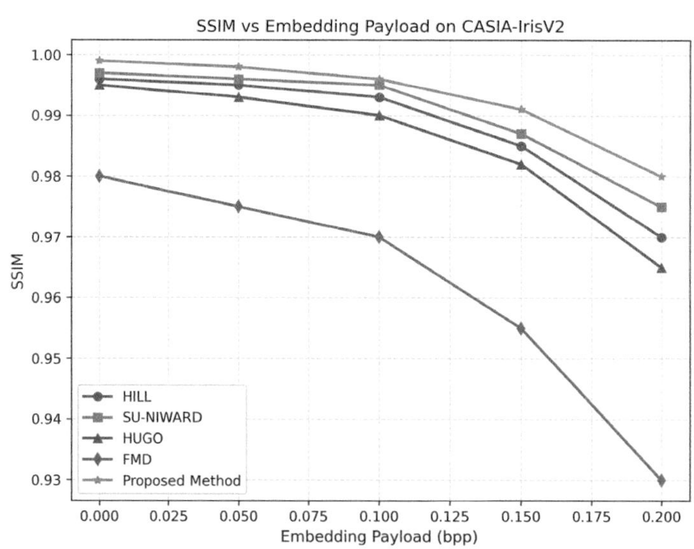

Figure 12.4 SSIM vs Payload.

MSE vs Payload: MSE (Mean Squared Error): Lower = better. Measures pixel-wise distortion. The graph shows MSE vs. embedding payload for various steganographic methods on the CASIA-IrisV2 dataset. As payload increases, MSE rises, indicating more image distortion. The Proposed Method maintains the lowest MSE across all payloads, reflecting better embedding quality. In contrast, EMD shows the highest distortion. Among existing methods, S-UNIWARD and HILL perform better than HUGO, but still lag behind the proposed approach (Figure 12.5).

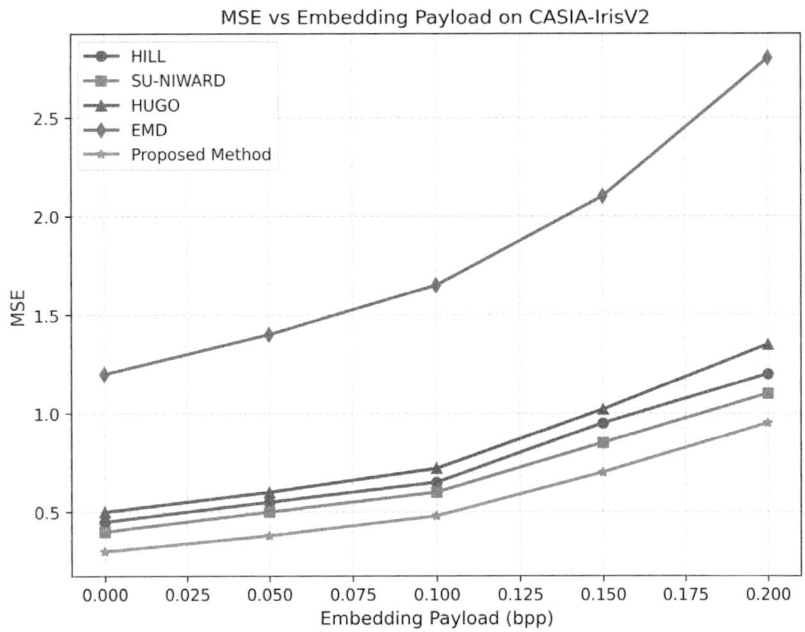

Figure 12.5 MSE Vs Payload.

MAE (Mean Absolute Error): Lower = better. Measures the average absolute difference between original and embedded image pixels. The graph shows MAE vs. embedding payload for different steganographic methods on the CASIA-IrisV2 dataset. MAE measures average pixel-level error; lower is better. As payload increases, MAE rises for all methods, indicating more distortion. The Proposed Method consistently shows the lowest MAE, while EMD has the highest, reflecting greater visual degradation (Figure 12.6).

Payload Capacity vs Recognition Accuracy: Assuming a bpp-to-kilobyte mapping for a 640×480 grayscale image (307,200 pixels), the graph shows Recognition Accuracy vs. Embedding Payload for various methods on the CASIA-IrisV2 dataset. As payload increases, accuracy generally drops, indicating degraded biometric performance. The Proposed Method maintains the highest accuracy across all payloads, showing strong robustness. In contrast, EMD shows the steepest decline, highlighting its reduced reliability for biometric use (Figure 12.7).

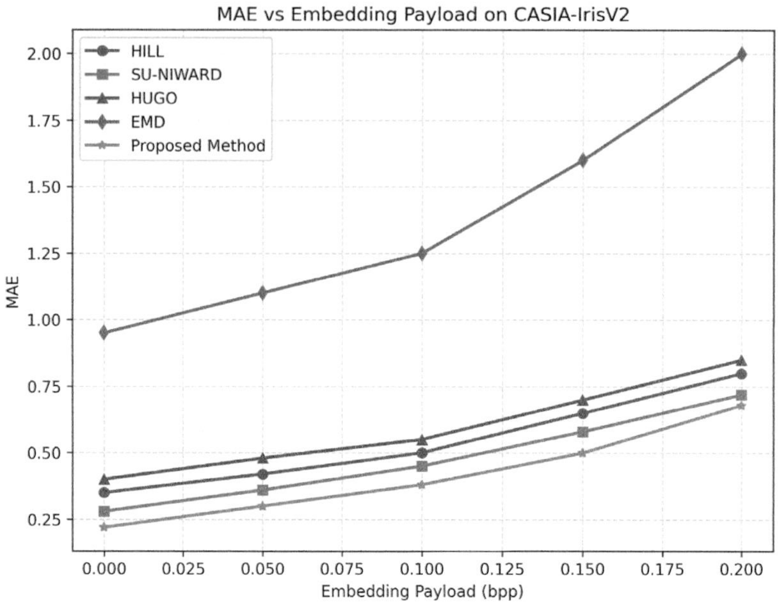

Figure 12.6 MAE vs Payload.

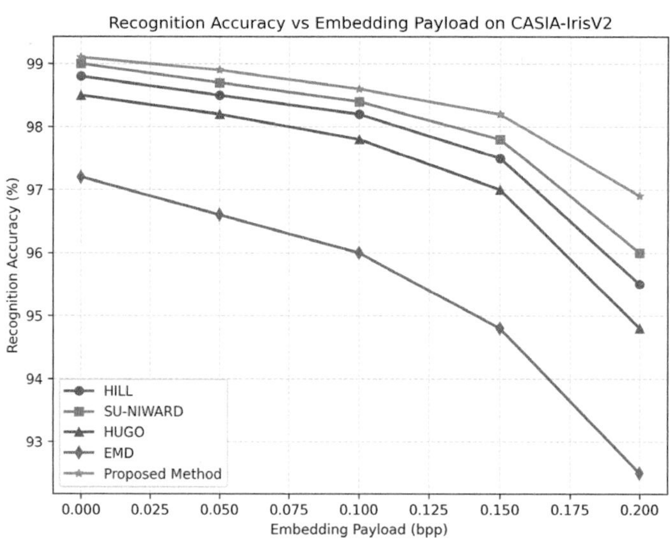

Figure 12.7 Payload Vs Accuracy.

FAR (False Acceptance Rate): Lower = better. Measures the risk of falsely accepting unauthorized users. The graph shows FAR vs. embedding payload for various methods on the CASIA-IrisV2 dataset. FAR measures the chance of wrongly accepting unauthorized users; the lower the better. As payload increases, FAR rises for all methods. The Proposed Method maintains the lowest FAR, ensuring strong

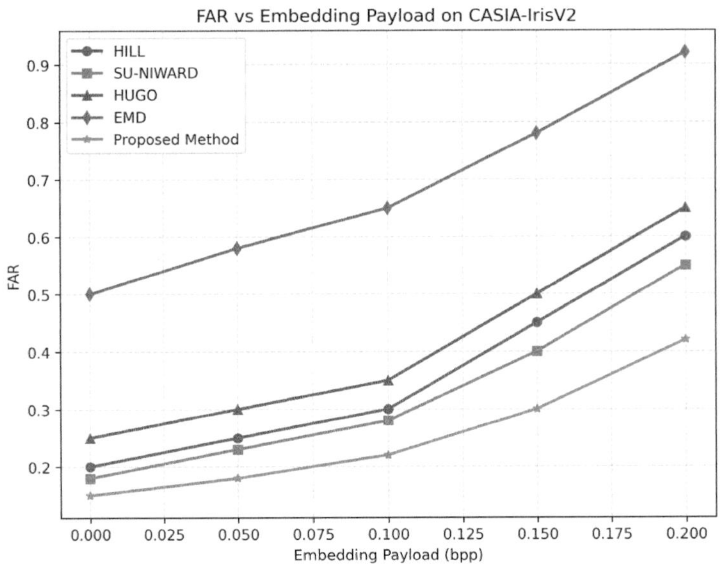

Figure 12.8 FAR vs Payload.

security, while EMD records the highest, indicating poor reliability for biometric systems (Figure 12.8).

FRR (False Rejection Rate): Lower = better. Measures the likelihood of rejecting valid users during authentication. The graph shows FRR vs. embedding payload for various methods on the CASIA-IrisV2 dataset. FRR measures the chance of wrongly rejecting authorized users; lower is better. As payload increases, FRR rises for all methods. The Proposed Method consistently shows the lowest FRR, ensuring reliable recognition, while EMD has the highest, indicating poor performance at higher payloads (Figure 12.9).

Justification: The proposed transformer-assisted reversible data embedding scheme represents a significant progress towards secure biometric processing. Transformers can extract the fine-grained texture of the iris by utilizing the attention mechanism globally, which is better than traditional methods. The application of Syndrome Trellis Codes (STC) for embedding schemes reduces distortion for disjoint hash embedding to ensure recognition performance, while hashing makes it secure for tampering detection and integrity. Full reversibility is achieved, by which the embedded data and the iris image can be losslessly retrieved. This incorporation of transformers, STC, and hashing in a single framework fits to be a robust solution to the existing security gaps, authentication, and robustness issues for applications such as biometric authentication, e-passports, etc.

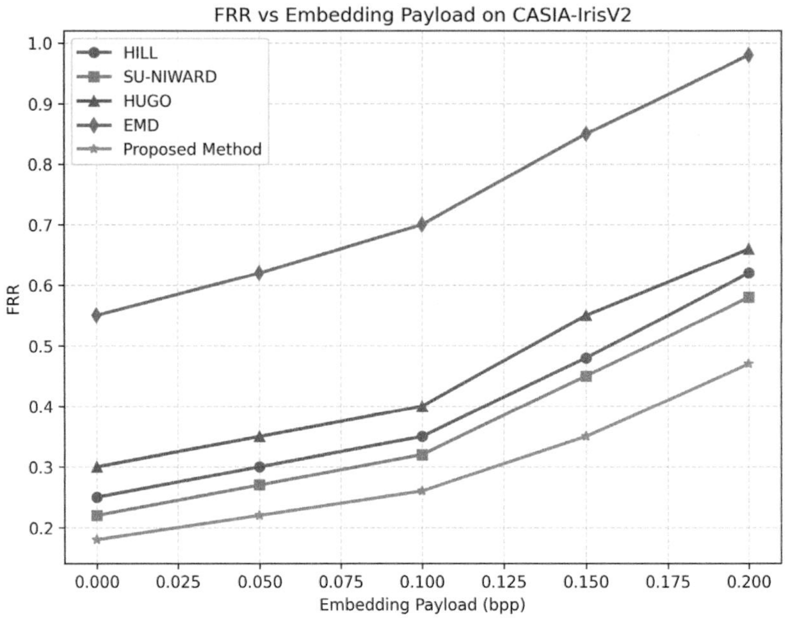

Figure 12.9 FRR vs Payload.

7. Conclusion and Future Work

To make decorrelated iris templates secure and secure them at the biometric integrity level, this work proposed a reversible data implantation mechanism for iris images using a transformer-based idea with data implantation through STC and hashing along with Syndrome-Trellis Codes. And calculated on CASIA-IrisV2 database in lower distortion (MSE and MAE), higher image quality (PSNR and SSIM), and GT concert (Accuracy, FAR, FRR) than HILL, S-UNIWARD, HUGO, and EMD. The results demonstrate the efficacy of our framework for secure and reversible biometric embedding. Further, in the next stage, multi-modal biometric fusion for enhanced security, lightweight transformer variants for edge posture, and adaptive solidity to ensure efficient storage will be areas of improvement. Quantum-inspired or chaotic techniques then enhance resilience embedding, and the embedded system disposition is used for real-time secure computing applications to secure the identity systems.

References

[1] Wang, C., Wang, H. and Li, J. (2020). Survey on reversible data hiding: techniques and applications. IEEE Access 8: 32178–32197.

[2] Ma, K., Zhang, W., Zhao, X., Yu, N. and Li, F. (2013). Reversible data hiding in encrypted images by reserving room before encryption. IEEE Trans. Inf. Forensics Security 8(3): 553–562, Mar. 2013.

[3] Wu, H.T. and Chen, J.H. (2015). Reversible image data hiding with contrast enhancement. IEEE Signal Process. Lett. 22(1): 81–85, Jan. 2015.

[4] Fallahpour, M. and Megías, D. (2013). Reversible and high-capacity data hiding in medical images. IET Image Process. 7(2): 105–113, Mar. 2013.
[5] Ni, Z., Shi, Y.-Q., Ansari, N. and Su, W. (2006). Reversible data hiding. IEEE Trans. Circuits Syst. Video Technol. 16(3): 354–362, Mar. 2006.
[6] Tian, J. (2003). Reversible data embedding using a difference expansion. IEEE Trans. Circuits Syst. Video Technol. 13(8): 890–896, Aug. 2003.
[7] Cayre, F., Fontaine, C. and Furon, T. (2005). Watermarking security: theory and practice. IEEE Trans. Signal Process. 53(10): 3976–3987, Oct. 2005.
[8] CASIA Iris Image Database, Version 2.0. Chinese Academy of Sciences Institute of Automation. [Online]. Available: http://biometrics.idealtest.org/.
[9] Wang, H., Qian, Z. and Zhang, X. (2020). Efficient reversible data hiding with context-aware pixel selection. IEEE Trans. Image Process. 29: 1749–1763.
[10] Mehta, A., Rathi, S. and Singh, L. (2022). Lightweight transformers for edge-based biometric processing. pp. 501–509. In Proc. CVPRW, Jun. 2022.
[11] Tai, W., Yeh, C. and Chang, C. (2009). Reversible data hiding based on histogram modification of pixel differences. IEEE Trans. Circuits Syst. Video Technol. 19(6): 906–910, Jun. 2009.
[12] Li, X., Yang, B. and Zeng, T. (2011). Efficient reversible watermarking based on adaptive prediction-error expansion and pixel selection. IEEE Trans. Image Process. 20(12): 3524–3533, Dec. 2011.
[13] Celik, M.U., Sharma, G., Tekalp, A.M. and Saber, E. (2005). Lossless generalized-LSB data embedding. IEEE Trans. Image Process. 14(2): 253–266, Feb. 2005.
[14] Celik, M.U., Sharma, G. and Tekalp, A.M. (2002). Reversible data hiding using gradient-adjusted prediction. In Proc. SPIE 5020: 1–10.
[15] Hong, W., Chen, T.-S. and Wu, H.-Y. (2012). An improved reversible data hiding in encrypted images using side match. IEEE Signal Process. Lett. 19(4): 199–202, Apr. 2012.
[16] Ni, Z., Shi, Y.-Q., Ansari, N. and Su, W. (2006). Reversible data hiding. IEEE Trans. Circuits Syst. Video Technol. 16(3): 354–362, Mar. 2006.
[17] Zeng, Y., Huang, J. and Shi, Y.-Q. (2009). Reliable reversible data hiding based on pixel differences. In Proc. MM, pp. 537–540.
[18] Fallahpour, M. and Megías, D. (2013). Reversible and high-capacity data hiding in medical images. IET Image Process. 7(2): 105–113, Mar. 2013.
[19] Zhang, X. and Ren, Y. (2017). Reversible data hiding in encrypted images by data hiding before encryption. J. Syst. Softw. 127: 192–202, Mar. 2017.
[20] Huang, H., Hwang, Min-Shiang and Tseng, Lin-Yu. (2013). Reversible data hiding for high quality medical images. pp. 1427–1430. In Proc. IEEE EMBC.
[21] Chen, Y., Shi, G. and Huang, J. (2013). Just noticeable distortion model and histograms modification based reversible data hiding. Inf. Sci. 221: 70–81, Mar. 2013.
[22] Khan, M. and Kumar, R. (2014). A block-based reversible watermarking technique for tamper detection and localization in medical images. Comput. Electr. Eng. 40(6): 1999–2010, Aug. 2014.
[23] Whitelam, C., Ross, A. and Crihalmeanu, M. (2015). Multimodal fusion of iris and periocular biometrics for cross-sensor recognition. Pattern Recognit. Lett. 57: 52–59, May 2015.
[24] Yalman, M. and Akar, A. (2010). An application of reversible data hiding based on R-weighted coding and neighbor-mean interpolation. Comput. Electr. Eng. 36(2): 282–294, Mar. 2010.
[25] Yalman, M. and Akar, A. (2014). Improved reversible data hiding using RCM and mean interpolation with edge-adaptive mechanisms. J. Vis. Commun. Image Represent. 25(8): 1890–1900, Nov. 2014.
[26] Rani, M.S. and Selvarani, S. (2024). Contemporary approaches on reversible data hiding in iris image using deep learning techniques. In Proc. IACIDS, Lavasa, India, Mar. 2024, doi: 10.4108/eai.23-11-2023.2343227.
[27] Song, B., Dongdong Zhao, Jiang Yan, Huanhuan Li and Hao Jiang. (2024). BioDeepHash: Mapping Biometrics into a Stable Code. *arXiv*, Aug. 2024.
[28] Wang, M., Bozkir, E. and Kasneci, E. (2025). Iris Style Transfer: Enhancing Iris Recognition with Style Features and Privacy Preservation through Neural Style Transfer. *arXiv*, Mar. 2025.

AI in Agriculture and Environment

Chapter 13

Paddy Disease Identification via Deep Learning

A ResNet-50 Approach

R. Gunavathi,[1,*] *S.B. Thamarai Selvi,*[2] *Shikalgar Nilofar*[3] and *Savithri M.*[4]

1. Introduction

1.1 Paddy Disease Classification

Agriculture, which is the pillar of global food availability, is constantly beset by crop diseases that significantly reduce harvest yields. With the onset of modern technologies such as Convolutional Neural Networks (CNNs) within the recent past, agricultural production has been revolutionized at an unparalleled rate of development. These technologies enable the early detection and precise identification of plant diseases, such that these damages can be prevented prior to widespread effects. One of the earliest applications of the methods is the identification of rice plant diseases such as Brown Spot, Blast, and others. With the application of CNNs, crop producers and crop specialists can promptly detect and respond to these pests. The eventual result is the timely, fact-based intervention that not only enhances the productive capability of crops but also improves sustainable production methods. Overall, the injection of high-technology solutions such as these ensures that producers' decisions within agriculture become more effective, fact-based, and robust against disease invasions.

[1] Associate Professor, School of Sciences, Christ University, Bangaluru, India.
[2] Assistant Professor of Computer Science and Engineering, K S Rangasamy College of Technology, Tiruchengode, India.
[3] Research Scholar, School of Sciences, Christ University, Bangaluru, India.
[4] Assistant Professor, School of Sciences, Christ University, Bangaluru, India.
* Corresponding author: gunavathi.r@christuniversity.in

DOI: 10.1201/9781003635178-18

Fitting on top of a robust dataset is the first step when it comes to tackling the issue of paddy disease marking. This type of dataset has a vast range of paddy disease images, which are carefully annotated with the related disease type. One of the reasons why a carefully maintained dataset is crucial is that it is a foundation on top of which robust deep learning plans are correctly trained. Data preprocessing plans are then used when it is about normalization and boosting of such a dataset, so that such a model is at a point where it is able to effortlessly acquire knowledge based on diverse data while generalizing appropriately in unknown data.

Selection of a correct deep learning model is of paramount importance when seeking to come up with a rightful disease classifier of paddy diseases. Because of its enhanced performance when it comes to image classification, a common choice is making use of a foundation in the form of ResNet-50. On top of its inherent ability to have a deep framework and skip connections, it is effective when it comes to extracting fine features and patterns in photographs of diseases of paddy plants. Although the use of the model is not done in a vanilla state per se, it is fine-tuned so as to fit within specified attributes within a paddy disease dataset. Here, parameters and even hyperparameters are fine-tuned to fit within a specified number of diverse diseases within a dataset, so that, in effect, it becomes an expert when it comes to discriminating between several diseases. Training of deep learning is stepwise and gradual, where a network is constantly being fed pictures with captions. Gradually, it allows learning so as to enhance internal representation, such that gradually stepwise increased accuracy is realized when diagnosing diseases within paddy plants. Test of performance and rather than rather being overfit, a part within a dataset is preserved within validation. Such a part within validation serves within a test function so that a model works sufficiently with a new piece of information it has never seen before. At its peak is a deep learning model which is trained with the inherent ability of fine designation of disease within paddy plants within hitherto images. The implication of such a model is that it is of crucial use within agriculture. Because of its ability to perform quick disease diagnosis, it is comparable to the said diagnosis, enabling the taking of appropriate corrective measures to avert the spread of the diseases and minimize losses within the crop plants. Such contributions in a technical form lead directly to increased crop yield, while in the long term, they support a sustainable approach to crop development.

Figure 13.1 displays paddy leaves with a representation of healthy and disease conditions. Images of these give a summation of a range of situations that are observed in the field. On the right-hand side and on the left-hand side, although in reference to the clean, unaffected status, a representation of healthy leaves is shown. Leaves of disease such as brown spot, rice blast, healthy leaf, and bacterial leaf blight are represented in order to have an idea about their distinctive features and pattern of occurrences [29]. This powerful comparison of diseased and healthy specimens is a sign of the requirement for proper and effective methodology for the classification of diseases. Sample images serve as the foundation for training deep learning models, enabling them to capture the inherent complexities involved in paddy disease classification. Beyond its practical application, this process exemplifies how artificial intelligence (AI) can be harnessed to tackle pressing challenges in

Figure 13.1 Leaf Images of Paddy Plants.

agriculture. The integration of modern technology with farming practices not only streamlines workflows but also showcases the creativity and versatility of AI in addressing real-world issues.

When examined closely, paddy disease classification through deep learning illustrates a paradigm shift: centuries-old agricultural practices are being reshaped by cutting-edge technology, highlighting its transformative power in ushering in an era of precision agriculture. With farmers and agronomists as the primary beneficiaries, this study seeks to design an accurate and reliable system for rice disease detection. Specifically, it employs a ResNet-50-inspired Convolutional Neural Network, aiming to achieve a target accuracy of 97%.

1.2 Few-Shot Learning

Few-shot learning is one revolutionary notion of artificial intelligence (AI), as it gives up the conventional need for large sets of tagged data used by the model for training. Artificial intelligence models almost exclusively asked for large sets of tagged information where one aspired for refined knowledge, besides some classes/work/responsibilities. But few-shot learning appears as an exception to the conventional trend of providing the model in the learning of knowledge on the basis of small samples of a class. Such a paradigm shift appears as of extreme importance within use cases of scarce available information, unknown classes/where it is too economic/feasible to accumulate large sets of tagged datasets. Few-shot learning [2] at the grassroots level breaks barriers within learning, allowing the model within the learning of knowledge extrinsically and generalize on the basis of very few samples. While the utilization of large sets of tagged information could be availed of, such utilization of the model shines by adapting within faster velocities on the basis of novel tasks/classes by unwinding salient representations and patterns based on small samples within the basis of the training base [27]. Meta-learning is such a sublime methodology within few-shot learning, constituted of training exclusively within diverse, small-scale learning jobs with just scarce samples within a class. With training operations of the kind, the model's triumph lies in quickly assigning knowledge whilst taking transferable representations assisting quick adaptation on the basis of novel, unseen tasks. Such a paradigm shift appears as of extreme importance within use cases within computer vision jobs, natural processing jobs, health application jobs, and many more such jobs. In computer vision, i.e., few-shot learning appears as of extreme importance where it is of the prime priority to detect novel object classes, but it is not practicable to have large, tagged datasets of those classes. With few samples, such jobs can detect and discern patterns and thus can detect and annotate novel objects with very high precision with no recourse to large tagged datasets [28]. Likewise, few-shot learning techniques have immense potential within applications like text tagging, sentiment tagging, or translation, where the fastest adaptation is of the ultimate importance within novel categories or novel languages. Again, within the medical or healthcare application domain, where access might be limited due to factors such as preservation of the patients' privacy or relative rarity of occurrence, few-shot learning can make medical tagging of the images, disease prediction, or drug development possible by allowing a model to learn with a few medical images or a few patient records. As a concept of few-shot learning is not merely a solution of evading the issues within the inadequacy of data but is of itself a foundation-laying concept towards AM's flexibility and generalizability within humongous novel-world diversity, it is also in sync with the spirit of the human learning which is such that a human can very soon learn a novel concept or a novel task with the exposure of meager size. Making it possible that the machines can very efficiently learn knowledge via a few examples, few-shot learning provides a foundation of AM-enabled application domains that can have quick adaptation, generalizability, and excel where customary methods do not work on the basis of confinements within the available datasets.

2. Related Work

Serge Savary et al. [3] comment on the global impact of insects and microorganisms attacking crops and causing significant losses. The paper points out the crucial role of these menaces in shaking the food system and impacting production, accessibility, distribution, and profitability of other parts of the economy. Shivani Sood et al. [4] propose a refined method of early wheat rust disease detection and achieve a remarkable 99.07% accuracy with the aid of VGG16. The paper proposes a transformative impact on agriculture and points out prospective future applications of refined feature detection strategies. Charles Farber et al. [5] comment on the transformative impact of expertise-level diagnosis using non-invasive spectroscopic technologies such as Remote Seeing (RS) and IR spectroscopy. They can save up to 50% of global agricultural output by overcoming the prevailing limitations in detecting plant diseases. They are yet a cost-competitive option when deployed in the field. Yanfeng Zhao et al. [6] utilized Double Generative Adversarial Networks (Double GAN) and bridged dataset imbalances of plant leaf diseases and attained 99.80% accuracy on the identification of plants and species and 99.53% accuracy on the identification of diseases. This promises advancing high-resolution image synthesis and robust models of the classification of diseases. YAQING WANG et al. [7] embody the strength of AI when data is abundant and acknowledge weaknesses when datasets are small. They introduce Few-Shot Learning (FSL) as a strategic corrective and critically investigate its conceptual leaning and demarcation from related issues and experimental vagueness. Vivek Kaul et al. [8] inquired about the revolutionary impact of deep learning on medical practice and gastroenterological and endoscopy practice and gave a brief account of the historical achievement of breakthroughs made on the front of anticipating and assessing risks and diagnostic processes, and streamlining of work processes. Imran Haider Khan et al. [9] recount a revolutionary hyperspectral image-based AI approach that detects early wheat diseases with an overall accuracy of greater than 82.35% and proposes transformative change on the agricultural front.

Mohsen Azadbakht et al. [10] proposed a precise detection of wheat leaf rust using spectral data and ν-support vector regression with higher performance, particularly under low Leaf Area Index situations, and with a high R2 value of nearly 0.99. Parisi et al. [11] assess lifelong learning problems for artificial systems and survey neural network approaches to prevent catastrophic forgetting, and identify the need to do more work and achieve robust lifelong learning of autonomous agents and robots based on biological system inspirations. He et al. [12] propose a new methodology called residual learning and suggest it for training deep neural networks. This had better optimization and accuracy, and outperformed the winners of the ILSVRC 2015 classification task. Further, it recorded remarkable improvements in visual understanding tasks and object detection using COCO. Hussain et al. [13] employ deep learning, i.e., Convolutional Neural Networks (CNNs), and utilized it to classify wheat diseases with 84.54% accuracy using in-place images and created a useful tool utilized by farmers to avoid losses and rescue crops. Manohar et al. [14] employ image processing and instantly detect and classify bacterial leaf blight, brown spot, and leaf smut in rice plants, and hope to strengthen the detection of diseases and control. Islam et al. [15] explore autonomous detection of paddy leaf diseases of Bangladesh using deep learning CNN models, and with Inception-ResNet-V2 recorded 92.68% accuracy, and can work and substitute faster than manual conventional works. Senan et al. [16] utilized image processing and a five-layer CNN and achieved 93% accuracy in the detection and typing of paddy diseases, and outperformed existing methods. Upadhyay et al. [17] use images and a fully connected CNN and can detect early-stage brown spot disease of rice plants, properly achieving a high accuracy of 99.20%, and surpassing other methods using CNNs. Simonyan et al. [18] explore the influence of CNN depth on image recognition and came in first and second places at the ImageNet Challenge 2014, and offer highly performing ConvNet models, which are of interest and use when performing computer vision tasks. Krithika et al. [19] suggest automation of salad cucumber leaf disease detection for intelligent organic farming in India using K-means clustering and the Support Vector Machine algorithm.

Lamba et al. [20] suggest an AI-based disease classifier of paddy plants treating fungal and bacterial leaf diseases with a new architecture of CNN, dataset augmentation based on GANs, and achieved a very high 98.23% accuracy of classification. Anami et al. [21] designed a color-based computer vision system that autonomously detects and classifies different forms of stress in paddy crops. They achieved 89.12% accuracy with BPNN, 84.44% accuracy with SVM, and 76.34% accuracy with k-NN. This system can help ensure farm-related decisions. Bhimavarapu [22] is about meeting the increasing demand of rice and taking urgent attention to detecting leaf infections. Utilizing Convolutional Neural Networks (CNN) and a new Improved Activation and Optimizer Function (IAOF), the work improves our ability to predict rice diseases with higher accuracy than previous works on efficient identification. Su et al. [23] assess new methods based on deep convolutional neural networks on smartphones for accurate and timely identification of dominant rice diseases. They achieved nearly 81.87% based on training data and 81.25% based on validation data with a 120-image dataset. Latif et al. [24] announce a highly accurate system of DCNN-based transfer learning with 96.08% average accuracy of identification of six different leaf diseases of rice and far surpasses existing works of literature on the work of disease classification. Senthil Pandi et al. [25] investigate deep learning detection of plant diseases. They introduced Dilated Convolutional Neural Network (DCNN) with Global Average Pooling (GAP) and made the detection process more efficient and accurate. Kusumo et al. [26] employ machine learning using RGB color features and detect corn diseases very significantly, and introduce a promising approach toward minimising economic loss due to outbreaks.

2.1 Problem Statement

In disease classification of wheat, the inclusion of the few-shot learning method [1] of the EfficientNet is a tried-and-true remedy for avoiding major challenges in disease identification within wheat plants. Application of the emergent methodology provides promising results of effectiveness, with a superior ability to notice and differentiate diseases through a mind-boggling level of success of 93.19%. Observant is the achievement of such a level of success being reached regardless of the constraint of dataset size, which provides a level of optimality of the model and adjustability by virtue of fewer illustrations. Such a breakthrough is built on the inclusion of the EfficientNet model, realization of a blinding level of accuracy of almost up to 98.5% within the CGIAR Dataset. Such effectiveness is more than an exhibition of sound effectiveness of the model but more a presentation of computational optimality when dealing with limitations of data and resources within disease detection of diseases within deep learning paradigms. Such a superior level of effectiveness, especially when dealing with a few training images, is a demonstration of robustness when dealing with limitations within the data used in disease detection. In practice, major challenges are efficiently nullified by the system, with the entire resolutions envisaged within major issues:

- Data Limitation Challenge
- Computational and Resource Constraints
- Adaptability and Generalization
- Constant Retraining Burden

Hints of the prospective result of precision do not end with the methodology itself, but also of a sustainable alternative methodology of disease identification of the wheat crop. By minimizing iterative returning of the model towards the new disease classes, it diminishes computational heftiness, one step further to the active crop industry disease identification model. But if the excellence of the methodology is to be fully proved, effortful comparative research against dominant models becomes imperative. Such research will not only authenticate the proposed approach but will equally place emphasis on its values and transformational ability within the domain of wheat disease classification.

3. Methodology

Such a system is a paradigm shift in the characterization of paddy diseases using a state-of-the-art deep learning approach for real-time and precise distinction of a few disorders, and therefore is a very high-value addition for farmers. As a CNN-motivated system with regard to architecture per se, using the architecture of ResNet-50, which is very notoriously renowned for very exceptional performance placed by it by way of use in image classification problems (see Figure 13.2). While adapting such a successful architecture toward its very special need of identification of paddy diseases, we make a few adaptations, such as adjusting the final classification layer of a model so that it is feasible to have a few numbers of a kind of disease within a dataset. Making a model efficient and robust requires painstakingly compiling a humongous array of images simply depicting a few numbers of infections of paddy carefully. Each of these images is painstakingly correctly annotated based on a few numbers of a kind of disease. All these are directed toward the accuracy improvement of a model and its generalizability under changing conditions. Standardization and augmentation strategies are used to standardize and augment a dataset toward considering variations of the external environment and quality of images. High-end augmentations are included within a system to maximize the ability of a model toward learning using very few training data available, significantly overcoming few-shot learning issues. Strategic processes of the kind of transfer learning and fine-tuning are also used to take advantage of pre-learned models, very specifically tailored toward use within the realm of classifiers of paddy diseases. During training iterations of a system, samples of a dataset are divided away towards validation to test the model's performance at every step and keep a check on overfitting.

The next trained model is a powerful asset that can be used to predict paddy diseases on unseen and novel photographs with instant and accurate diagnosis. Such a capability can be of immense empowerment to crop specialists and farmers with effective decision-making on crop health and disease control. Overall, the framework advocated is an optimized procedure that takes advantage of the deep learning quality so as to boost paddy disease discrimination and thus play a pivotal role in more sustainable and effective crop production. Its future impact is immense, with the promise of making a profound impact on reducing crop loss, increasing production, and easing the meeting of food security milestones at the international level. Such innovative methodology is at the cutting-edge within so far technological innovation is related and yet holds the potential of inducing a paradigm shift in what

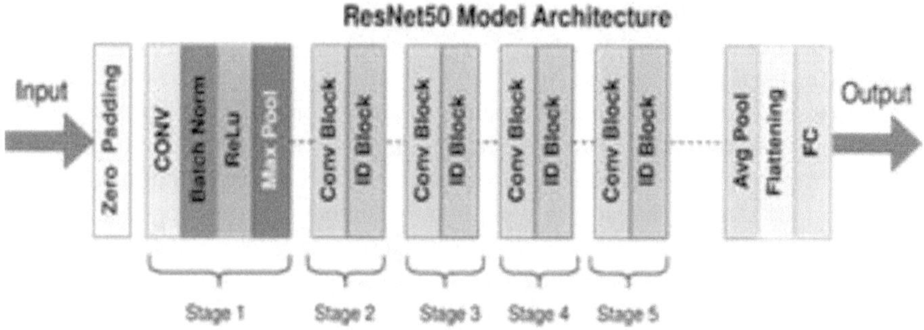

Figure 13.2 Architecture of the ResNet-50 Model.

we are capable of accomplishing for concerns within the crop sector. Convolutional Neural Networks (CNNs) emerge as a prospective candidate specifically tailored for tasks within image classification. Their strength lies in their ability to meticulously deliberate on complex details within photographs, such as edges, forms, and texture, allowing effective classification. ResNet-50 is a very effective candidate within a vast array of CNN constructions, mirroring a vast array of effective strong performances within a multitude of tasks and excelling to a remarkable extent within image classification. ResNet-50, or simply termed as Residual Network with 50 layers, orients such layers within four clear-cut blocks that are made up of successive convolution layers and then maximum pooling.

The design of architecture enables the network to take a gradual hold of fine details of images and detect refined patterns at high efficiency for effective classification. At the final end of the network, a conspicuous component by the name of a fully connected layer makes the final call. It decides within which class an image is. The flexibility of ResNet-50 has been shown along a sequence of applications for the classification of images, such that it is a reliable option. It can be applied perfectly even in the core area of classification of paddy diseases, which is ever generating output of superior quality. Taking advantage of the inherent strength of CNNs, particularly of the type of ResNet-50, is a bright future path to overcome those thorny issues of classification of paddy diseases. One of the beaming qualities of ResNet-50 is the successful functioning even in conditions where a variety of qualities and conditions are embodied by the images. Such a kind of versatility is ever more important within the scope of the area of the application of classification of diseases of paddy, where the conditions of the environment and quality of an image greatly vary. The ability of such a network to detect the features and figures automatically and ignore those variations is a prime quality of the endeavour of quick and accurate identification of diseases.

Figure 13.3 is a system architecture diagrammatic map of a system showing how the various components of the system operate and interact. It is a diagrammatic presentation made as a reference when explaining how a system works. By showing, through a diagrammatic presentation, how the various components of a system interact with each other, the figure presents a conceptual and transparent picture of

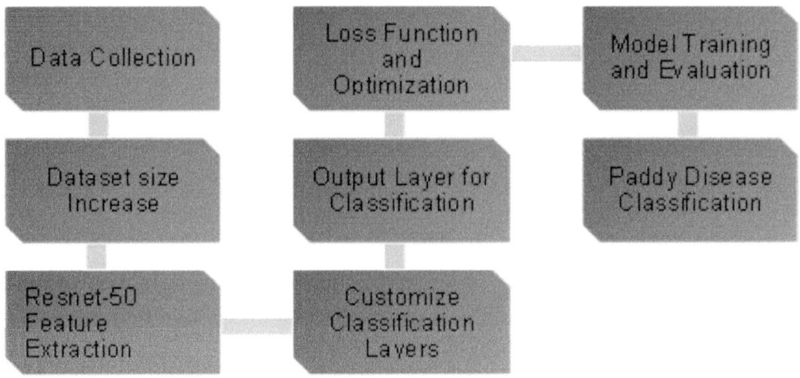

Figure 13.3 ResNet-50-based paddy disease classification architecture.

how a system works. It is not very detailed technically, such that it is simple for the user to understand how the many different components of the system work together. Overall, Figure 13.3 is a diagrammatic presentation that gives a basic representation of the composition of a system and of how the main constituents of the system work together.

Data Collection

It covers step one, where a full dataset of images of diverse paddy diseases is obtained such that every one of the images is adequately annotated with the relevant type of disease. This type of dataset is used as the foundation on which training and testing of the new system are based, such that it is put under an enormous spectrum of examples so that it can be adequately trained. Every one of the images is diligently executed such that the system is trained based on a diverse spectrum of examples of diverse diseases that occur on paddy plants. This type of dataset forms the foundation on which the system is trained, such that it can identify and classify between an enormous spectrum of types of diseases. This type of high-end dataset renders the system perfectly prepared to identify and classify an enormous spectrum of paddy diseases adequately, such that farmers and agriculturalists can tackle crop health deeply.

Dataset Size Increase

For boosting the capacity of the model to acquire knowledge, data augmentation processes are used, which essentially augment the size of the dataset virtually using rotation, flipping, and scaling techniques. Through this method, the model is prevented from overemphasizing certain features of the data, making it better suited to acquiring new knowledge it is not aware of. Other than this, all the images are resized at a constant size, easiest for alignment purposes, which is a 224 × 224 pixel size commonly, in conformance with the input size anticipated by the model ResNet-50. Pixel values are also normalized within a defined range, like [0, 1] or [–1, 1], such that the data converges better using the model. Data enlargement

Table 13.1 Dataset Images Specification.

Format	JPG
Size	6–7 MB
Dimension	224*224
Horizontal Resolution	96 dpi
Vertical Resolution	96 dpi

processes like rotation, scaling, flipping, and even random cropping are further used in introducing added diversity in favor of better robustness and performance of the model.

Table 13.1 refers to some features and attributes of a specific dataset of images of a specific application, such as the training of a model. These specifications refer to important information such as a specific number of images, their size, and the scope of subjects where they span so as to clearly identify a composition of a dataset and its application to a specific task.

Resnet 50 Feature Extraction

Pre-trained ResNet-50 is utilized in the classification of paddy diseases. We first load a pre-trained ResNet-50 model, except for its final classification layer, so that its layers remain frozen to retain learned features. It inhibits any upgradation of the parameters of our model used in the training step. We input pre-processed pre-learnt images of paddy diseases into pre-trained ResNet-50 model layers. It absorbs fine patterns along with features in the input images and extracts high-level conceptual features. It is crucial for effective feature extraction in leveraging knowledge used in fine feature detection by ResNet-50. We derive features from mid-level layers so that we can produce feature vectors representing prominent features in the images. It optimizes the model's ability to perceive as well as classify paddy diseases effectively. Pre-training in a very extensive dataset, such as ImageNet, optimizes the model's ability to perceive fine features in our special use case, such as images of paddy diseases so as to optimize the effectiveness used in feature extraction usable in the next step of classification.

Figure 13.4 is a series of photographs exemplifying the active healthiness of paddy plants. They are photographs exemplifying the strong and active status of unaffected paddy plants and are used as a comparison standard of ideal status. These exhibit full green cover, regular growth habits, and the absence of any witnessed signs of disease. Such visual records in pictorial forms of active paddy plants are a handy standard of comparison that permits easier discrimination when lined up against diseased plants' photographs. These variable set specimens of active vegetation provide a standard of comparison, providing accurate classification and identification of the disease formed on paddy plants by visual comparison. These provide handy materials providing training of models permitting sharp discrimination, providing deep acquaintance with various features developed on active paddy vegetation.

Figure 13.4 Images of Healthy Paddy Crops.

Customize Classification Layers

Similarly, when fine-tuning of ResNet-50 is carried out to achieve effective classification of paddy diseases, the last classification layers are fine-tuned. These fine-tunings prove imperative in restructuring the architecture suitably, utilizing the respective number of disease idiosyncrasies of the given dataset. Personalization is required when making the model independently distinguishable to differentiate various types of paddy diseases. Tweaking makes the model personally fine-tuned with regard to considerations of the dataset idiosyncrasies, so it is appropriately well-suited to discern and distinguish the voluminous plethora of diseases of paddy plants. With expert fine-tuning of a conclusive utilization of paddy disease classification, the learnability of a given model is optimized such that it is faster and efficiently makes effective classification. Such a thoughtfully crafted framing enhances the usability of the model, allowing identification and management within paddy diseases of farm fields.

$$C = CustomizedLayers(F_i) \tag{1}$$

Customized Layers function (C) used for paddy disease classification customizes features (F_i) identified using the aid of ResNet-50 are specified by Equation (1). It is required to change the architecture of ResNet-50 by introducing customized layers so that it can fine-tune features learned for discriminative classification based on certain attributes of paddy disease. This step fine-tunes a model representation so that it can bring out a meaningful manifestation of the pattern for discriminative paddy health. Customization allows features identified using the aid of ResNet-50 to be integrated very simply in a customized classification system based on subtleties of paddy disease discrimination.

Output Layer for Classification

Final classifications are generated by the output layer. Its parameters are calibrated so that it maps equally onto classes of diseases within a set that have individual nodes within each of the diseases. For generating classifications, the layer applies activation functions like SoftMax that calculate probability scores of every one of the disease classes. They are representatives of some confidence of an input image within a model being within a specific disease class. Direct one-to-one representation is the number of nodes representing unique diseases learned by a model when generating classifications. Through the use of softmax activation, such probabilities are calculated by the layer so that correct classification of input images is generated within the target classes of diseases. It is simply the final decision-maker of interpreting knowledge learned by a model within meaningful terms when constructing effective detection of diseases.

$$O_i = Softmax(C) \tag{2}$$

Softmax function $O_i = Softmax(C)$ is used for the conversion of customized classification layer C into a probability distribution specified by Equation (2). It holds the model's normalized output, which has a probability of each of the paddy disease classes. Then, the maximum probability class of a probability distribution O_i is selected as the outputted paddy disease in order to ensure proper classification at model inference.

Loss Function and Optimization

During model training, performance is rated using a proper metric termed a loss function, whose use is in categorical cross-entropy where many classes are involved. This type of loss function is a utility when measuring how effective a model is at predicting against actual values in a bid to maximize general accuracy. Then, optimization is put into context using algorithms such as stochastic gradient descent or Adam in a bid to optimize a selected loss function. Optimization improves the parameters of a model such that it evolves step by step to improve performance. This type of step-by-step improvement is significant in ensuring a model arrives at a point where it is optimal, such that its prediction aligns with actual values within a training dataset. Proper selection of a loss function and selecting an optimization algorithm is of immense significance in selecting a model with better ability to predict paddy diseases correctly. There is a need to maximize the prediction of a model within a span of time, such that it minimizes the gap between what it predicts and what happens in real life. This makes a model more effective practically.

Figure 13.5 encompasses integrated visual databases of several diseases and infestations of rice plants. These photographs illustrate specific appearances and appearances of some of these diseases, like Brown Spot, Blast, and Bacterial Leaf Blight, and other insects. They remain a significant reference tool for farmers and agronomists' rapid identification of the top problems they encounter in their rice crops.

Figure 13.5 Images of Unhealthy Paddy Crops.

Model Training and Evaluation

When training is practiced on a model, it is trained on a specially prepared set. The set is divided into a training set, used for training a model, and a validation set, about which a model's performance is checked. When a model is acquiring knowledge during training, it is always modifying its internal configurations, or parameters, by an optimization approach chosen for its ability. Through iteration, a model may enhance its knowledge about a set. In order to prevent letting a model overemphasize parts of information it was trained on and yet perform tolerably when shown new information, it is periodically checked using a validation set. Frequently evaluating it this way lets you check a model's performance with regard to information it had not yet encountered. It is a way of checking if a model did not simply memorize information when training was taking place, but learned information. Evading overfitting is a matter of constant scrutiny. Overfitting is a serious concern where a model attentively attends too closely to training set details at an undesirable price of its ability to tolerate information that has not seen at all in real life. Keeping track of a model's performance based on a validation set balances training based on training information and with the ability to respond adequately when shown new information, so a model remains effective when used in real-life scenarios outside training.

$$\theta^* = \arg\min_\theta \frac{1}{N} \sum_{i=1}^{N} L(O_i, y_i) \tag{3}$$

Equation (3) is utilized when optimizing the training of models. It is utilized when deriving optimum values of θ^* so that parameters are optimized such that a loss (L) of a cross-entropy is reduced on average. It is minimized as a necessity when training a strong ResNet-50 model when performing effective classification of paddy diseases.

Paddy Disease Classification

Having moved beyond its training and test stages, the model is now prepared for actual use in the detection of paddy disease. When new images of damaged crops of paddy are presented to it, in its trained form, it utilizes learned patterns and attribute use for predicting the likelihood of disease categories. Through such a feature, the detection of diseases is facilitated and made faster. For farmers and agricultural experts, such a feature means that they have at their disposal a useful tool. On the basis of such results of prediction, control of diseases and ensuring the paddy crops remain healthy are sufficiently determined. Through such results of prediction as a guide toward the possibility of action, an early attempt at control of disease development is facilitated while maximum yields are attained. Through utilization within real-life contexts, it is quite clear how useful such a model is as a tool in assisting with crop activity. Through its detection of disease features, not only is control of disease facilitated and made fast and accurate in its development, but a lot goes toward maintaining a general healthy and productive status of paddy crops at large, toward a successful crop food production activity.

$$y_{pred} = argmax(O_i) \tag{4}$$

Arg max is used for choosing the class of the paddy disease in y_{pred} by taking the maximum probability index from the output layer O_i. The arg max function from Equation (4) receives the maximum likelihood class out of multiple classes being predicted within a model. Arg max is a final choice of classification, choosing the maximum likely paddy disease for a specific input image.

4. Result and Discussion

A performance test of the classification system of paddy diseases registered stellar figures with a consistent accuracy of 97% gained by vigorous training and validation. Such stellar accuracy warrants a word of praise for the capability of such a system of successful identification and classification of some paddy diseases, which is of immense applicability for use in agricultural practice. The brilliant visualization of Figure 13.6 is an obvious summarisation of such a triumph of such a system. And of the highest priority, such stellar accuracy of a particular type of disease is overpowered by what is much more remarkable, the successful generalisation ability of such a system on an immense array of diseases and stages. Such a performance is of high priority for use in real agricultural practice since the co-existence of a plurality of diseases is an important factor. Such an ability to effectively manage multiple disease types has been both effective and reliable in actual decision-making processes of agriculture. Such accuracy, apart, is not only specific to mere identification but also to the spillover effects of applying such a practicum of actual on-the-ground decision-making processes in agriculture. Such a deciding factor of leeway towards such a triumph of such a system is within its fine-tuning procedure.

Table 13.2 is itself a pictorial roadmap that is a friendly guide toward the navigation of the procedure of classification of paddy diseases. It is a simple depiction of the way the system takes input information to discriminate between varying diseases. It entailed adjusting the learning rates of certain layers such that it led to significantly improved performance of the model. Making adjustments improved convergence such that the model attained the best performance within fewer iterations more

Table 13.2 Classification of Paddy Diseases.

S.NO	IMAGE_ID	LABEL
1	10010	Normal
2	10011	Bacterial_leaf_streak
3	10012	Hispa
4	10013	Normal
5	10014	Normal
6	10015	Tungro

quickly, and in parallel brought reduced training time. Practical adjusting brings out further the system's efficiency and application to useful agricultural practice fields. This table is a pictorial snapshot covering the model's efficiency in discriminating between varying types of diseases. It pictorially compresses crucial steps within the procedure of classification, such that it is easy for the user to comprehend in a friendly manner. The pictorial representation within Table 13.2 covers a systematic procedure within which the functioning of the model works toward portraying its ability to properly sense and discriminate variegated forms of diseases of paddies. Each column and each row within the table is representative of some parameters within the procedure of classification, such that it covers a systematic display of decision-making of the model. It is such a pictorial representation that is useful toward capturing within it the details of the procedure of classification, such that it portrays maximum effectiveness toward handling varying input information.

Figure 13.6 is a reference to a pictorial presentation or picture recorded particularly for diagnosing paddy diseases. It is simply a picture recorded particularly for the identification of potential issues within a rice plantation. This type of picture is very crucial within agriculture because it makes it easier for farmers and experts to access the health of paddy plants through a pictorial interpretation. Through features seen within a pattern within such a picture, such as color change, marks, or unusual growth, it is very easy to identify some of the features of some of the diseases, like

Figure 13.6. Paddy Disease Detection Image.

Table 13.3 Model Accuracy Comparison.

Epoch	Accuracy	Average Loss
1	92.4	0.213985
2	94.6	0.161165
3	95.7	0.185235
4	97	0.120766

brown spot, blast, or bacterial leaf blight. These pictures are very vital within training as well as within testing deep learning models, particularly developed for automatic identification and detection of some of the paddy diseases. Finally, a picture properly recorded, particularly for the identification of paddy disease, ensures early detection and particularly control of the disease within a crop for a healthier reproductive rice farm.

It is 97% accurate, as can be seen from Table 13.3. That is quite a remarkable achievement of farmers and agricultural practitioners. That level of accuracy is quite specific in that it can identify and diagnose the diseases of the paddies. That level of achievement is akin to having a very useful tool where farmers can identify the disease at its early stages and do something about it right away. As soon as such issues are discovered early, the farmers will be able to act on them right away and thus get more from their crop besides minimal losses. Such a scale of the success of a system is not just a victory of technology but a valuable resource for the entire nation practicing agriculture.

It announces the mechanisms whereby state-of-the-art technology with a veneer of advanced learning has great promise of giving a fillip toward farm woes. The system brings farmers a lift toward managing paddy diseases, a state-of-the-art remedy toward staples of yesteryear. It is a technologically advanced sidekick of the kind that is blessed with the ability to detect and categorize diseases within the fields so that farmers can initiate remedial measures regarding their crop care. At its very core, it is not such a triumph regarding technologies at all; it is regarding elevating farmers and turning around farming practices. Success along the lines of the system is a stride toward the right thing regarding the elimination of paddy diseases along sustainable yet effective lines. It is a salubrious portent regarding the future of the crops farmed along the lines of technology being the very cornerstone of health within crops and robust paybacks along the lines of the farmers and their people.

Figure 13.7 is a performance graph of a model of a ResNet-50 architecture. It is a graphical representation that shows two significant features: accuracy and loss. Accuracy is an indication of how correctly a model is able to predict the desired outcomes based on its performance. Loss is an indication of the level of error in a predictive function of a model – low loss is a sign of a correct prediction. There is a graph of both training and validation steps, which can ideally be observed carefully to assess a model's ability of learning. It is a strong tool for a researcher or a developer that shows a visual snapshot of a model's ability based on its ability to predict the correct results with the increment of training.

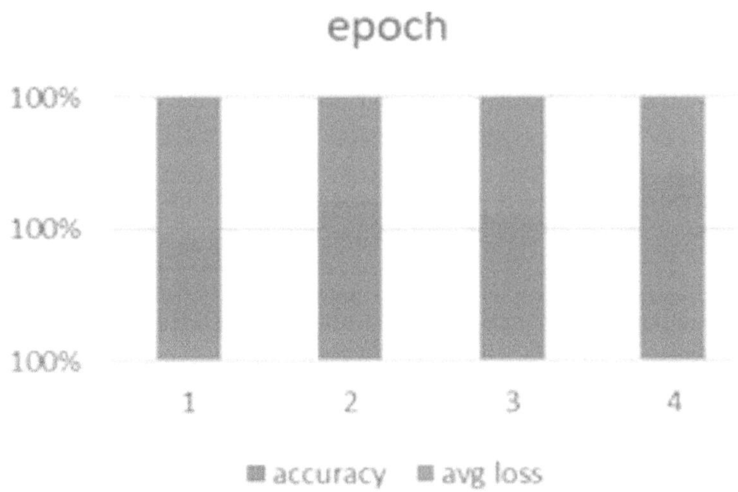

Figure 13.7 Graph of Model Accuracy.

5. Conclusion

Paddy disease detection system using a Convolutional Neural Network (CNN)-running ResNet-50 is a game-changer when it comes to agriculture. This kind of system is founded on complex algorithms on the computer, especially deep learning, which are fast and precise at detecting paddy plant diseases. Through careful uptake and formatting of information, fine-tuning it by preprocessing and complex manipulation of optimization of dataset-wise, such a system is learned based on limited training information input. Such a module is vital in the detection of diseases, which is a rare procedure because such conventional mechanisms would malfunction. Through such manipulation, such as in using transfer learning and fine-tuning, pre-gained knowledge is inherited for it to converge on such a technical procedure, such as in mixed paddy disease identification. Relentless training and stringent validation exercises on a sound dataset make such a system not only fast but dead-on target when it comes to output. Such dead-on target output guarantees relief to farmers and agriculturists on the basis of the adoption of effective control measures towards crop health management. At a broader scale, such a project is a guiding light towards such a transformative prospect, towards solving real issues, and getting artificial intelligence (AI) and deep learning. With improvement towards making the world have food safely and also a full-face lift towards restructuring agriculture on the basis of technology, such a project is a reference point on the basis of such a transformative prospect towards such a great issue-solving technology. Its prospect is not only on the basis of loss reduction on crop bases while maximizing crop output, but on the basis of restructuring agriculture on the basis of technology. With the ongoing evolution of such a system, it is on track towards a great contribution by making the world have a food supply while making sustainable mechanisms on the basis of crop output.

6. Future Direction

For a wealthier data set, the acquisition of broader paddy disease images is required, which improves the differentiation model performance of a variety of diseases. Geographical data and environmental variable integration enable the understanding of the disease pattern and its relation to corresponding regions and conditions. Structure optimization of the model has promise; investigation of a series of types of CNNs or the inclusion of further advanced models, such as attention mechanisms, can improve classification precision. Autonomous intervention mechanisms-based real-time disease identification has great promise applicable to agriculture. With compatibility with friendly smartphone apps or disease-detecting robots in the fields, farmers can take action early. Flexibility and scalability are requirements, including a sound system structure that allows integration of future disease classes or generalized use across a variety of crops.

References

[1] Alharbi, A., Khan, M.U.G. and Tayyaba, B. (2023). Wheat disease classification using continual learning. IEEE Access 11: 90016–90026. doi: 10.1109/ACCESS.3304358.

[2] Argüeso, D., Picon, A., Irusta, U., Medela, A., San-Emeterio, M.G., Bereciartua, A. et al. (2020). Few-shot learning approach for plant disease classification using images taken in the field. Comput Electron Agricult. 175: 105542.

[3] Savary, S., Willocquet, L., Lethbridge, S.J., Esker, P., McRoberts, N. and Nelson, A. (2020). The worldwide weight of microbes and nuisances on significant food crops. Nature Biol Evol. 3(3): 430–439.

[4] Sood, S. and Singh, H. (2020). An execution and examination of profound learning models for the identification of wheat rust sickness. pp. 341–347. In: Proc. Third Int. Conf. Intell. Maintain. Syst. (ICISS).

[5] Farber, C., Mahnke, M., Sanchez, L. and Kurouski, D. (2021). High level spectroscopic strategies for plant infection diagnostics. A survey. TrAC Trends Anal. Chem. 118: 43–49.

[6] Zhao, Y., Chen, Z., Gao, X., Melody, W., Xiong, Q., Hu, J. et al. (2022). Plant sickness recognition utilizing created leaves in light of DoubleGAN. IEEE/ACM Trans Comput Biol Bioinf. 19(3): 1817–1826.

[7] Wang, Y., Yao, Q., Kwok, J.T. and Ni, L.M. (2021). Summing up from a couple of models: A review on scarcely any shot learning. ACM Comput Overviews 53(3): 1–34.

[8] Kaul, V., Ensile, S. and Gross, S.A. (2020). History of man-made consciousness in medication. Gastrointest Endosc. 92(4): 807–812.

[9] Khan, I.H., Liu, H., Li, W., Cao, A., Wang, X., Liu, H. et al. (2021). Early identification of fine buildup illness and precise measurement of its seriousness utilizing hyperspectral pictures in wheat. Remote Sens. 13(18): 3612.

[10] Azadbakht, M., Ashourloo, D., Aghighi, H., Radiom, S. and Alimohammadi, A. (2020). Wheat leaf rust recognition at shelter scale under various LAI levels utilizing AI methods. Comput Electron Agric. 156: 119–128.

[11] Parisi, G.I., Kemker, R., Part, J.L., Kanan, C. and Wermter, S. (2019). Continual lifelong learning with neural networks: A review. Neural Netw. 113: 54–71.

[12] He, K., Zhang, X., Ren, S. and Sun, J. (2016). Deep residual learning for image recognition. pp. 770–778. *In*: Proceedings of the IEEE Conference on Computer Vision and Pattern Recognition (CVPR).

[13] Hussain, A., Ahmad, M., Mughal, I.A. and Ali, H. (2018). Automatic disease detection in wheat crop using convolution neural network. pp. 1–4. *In*: Proceedings of the 4th International Conference on Next Generation Computing.

[14] Manohar, N. and Gowda, K.J. (2020). Image processing system based identification and classification of leaf disease: a case study on paddy leaf. pp. 451–457. *In*: Proceedings of the International Conference on Electronics and Sustainable Communication Systems (ICESC).

[15] Islam, M.A., Rahman Shuvo, N., Shamsojjaman, M., Hasan, S., Hossain, S. and Khatun, T. (2021). An automated convolutional neural network based approach for paddy leaf disease detection. Int. J. Adv. Comput. Sci. Appl. 12(1): 280–288.

[16] Senan, N., Aamir, M., Ibrahim, R., Taujuddin, N.S.A.M. and Muda, W.H.N.W. (2020). An efficient convolutional neural network for paddy leaf disease and pest classification. Int. J. Adv. Comput. Sci. Appl. 11(7): 116–122.

[17] Upadhyay, S.K. and Kumar, A. (2021). Early-stage brown spot disease recognition in paddy using image processing and deep learning techniques. Trait Signal. 38(6): 1755–1766.

[18] Simonyan, K. and Zisserman, A. (2015). Very deep convolutional networks for large-scale image recognition. pp. 1–14. *In*: Proceedings of the 3rd International Conference on Learning Representations (ICLR); 2015 May 7–9; San Diego, CA.

[19] Krithika, P. and Veni, S. (2017). Leaf disease detection on cucumber leaves using multiclass support vector machine. pp. 1276–1281. *In*: Proceedings of the IEEE International Conference on Wireless Communications, Signal Processing and Networking (WiSPNET); 2017 Mar 22–24; Chennai, India.

[20] Lamba, S., Baliyan, A. and Kukreja, V. (2022). GAN-based image augmentation for increased CNN performance in Paddy leaf disease classification. pp. 2054–2059. *In*: Proceedings of the 2nd International Conference on Advance Computing and Innovative Technologies in Engineering (ICACITE); 2022; Location. Place of Publication: Publisher.

[21] Anami, B.S., Malvade, N.N. and Palaiah, S. (2020). Classification of yield affecting biotic and abiotic paddy crop stresses using field images. Inf. Process Agric. 7(2): 272–285.

[22] Bhimavarapu, U. (2023). Prediction and classification of rice leaves using the improved PSO clustering and improved CNN. Multimed Tools Appl.

[23] Su, N.T., Hung, P.D., Vinh, B.T. and Diep, V.T. (2022). Rice leaf disease classification using deep learning and target for mobile devices. pp. 136–148. *In*: Proceedings of the International Conference on Emerging Technologies and Intelligent Systems; 2022; Location. Place of publication: Publisher.

[24] Latif, G., Abdelhamid, S.E., Mallouhy, R.E., Alghazo, J. and Kazimi, Z.A. (2022). Deep learning utilization in agriculture: Detection of rice plant diseases using an improved CNN model. Plants 11(17): 2230.

[25] Senthil Pandi, S., Senthilselvi, A., Gitanjali, J., ArivuSelvan, K., Gopal, J. and Vellingiri, J. (2022). Rice plant disease classification using dilated convolutional neural network with global average pooling. Ecol. Modell. 474: 110166.

[26] Kusumo, B.S., Heryana, A., Mahendra, O. and Pardede, H.F. (2018). Machine learning-based for automatic detection of corn-plant diseases using image processing. pp. 93–97. *In*: Proceedings of the International Conference on Computer Control Informatics and its Applications (IC3INA); 2018; Location. Place of publication: Publisher.

[27] Azween Abdullah, S.. Nithya, M. Mary Shanthi Rani, S. Vijayalakshmi and Balamurugan Balusamy. (2023). Stacked LSTM and Kernel-PCA-based ensemble learning for cardiac arrhythmia classification. International Journal of Advanced Computer Science and Applications (IJACSA), 14(9): 2023.

[28] Thirumeninathan, V., Vijayalakshmi, S. and Palathara, S.T. (2024). Enhancing banana cultivation: disease identification through CNN and SVM analysis for optimal plant health. pp. 1–6. 2024 International Conference on Trends in Quantum Computing and Emerging Business Technologies, Pune, India, doi: 10.1109/TQCEBT59414.2024.10545264.

[29] Lekha, J. and Vijayalakshmi, S. (2024). Enhanced weed detection in sustainable agriculture: A You Only Look Once v7 and Internet of Things sensor approach for maximizing crop quality. Engineering Proceedings, Open Source Preview 82(1): 100.

Chapter 14

Plant Disease Detection Classification Model Using Segmentation and GLCM

Shalini Sharma,[1] *Rajesh E.,*[2,*] *Baskar Kasi*[3] and *Sakthi Govindaraju*[4]

1. Introduction

The economies of numerous impoverished nations are heavily dependent on agriculture. India's economy heavily relies on the agricultural industry. The agriculture sector of the Indian economy is responsible for providing food and other essential commodities. Hence, any damage to the crop results in a significant decrease in production, exerting a substantial impact on the nation's advancement. Plant leaves exhibit a high degree of fragility. Plant diseases frequently manifest as leaf browning or wilting. The dimensions and calibre of a crop are contingent upon the well-being of the plants responsible for its production. Hence, the early detection of leaf diseases is of utmost importance [1]. The expeditious containment and prevention of disease transmission within a farm yield superior outcomes. The detection of ailments in plants can be facilitated by examining the diversity and state of their leaves [2]. Successful implementation of this technique necessitates a level of talent and effort that is beyond the average. However, it would not be feasible on farms of such immense size. The identification of contamination within a plant might be facilitated by plant pathologists through the examination of distinctive indications, such as atypical markings, coloration, or shading [3]. The economies of numerous

[1] PG Scholar, School of Computer Science and Engineering, Galgotias University, India.
[2] Professor, School of Computer Science and Engineering, Galgotias University, India.
[3] Associate Professor, School of Computer Science and Engineering, Galgotias University, India.
[4] Professor, School of Computer Science and Engineering, Galgotias University, India.
Emails: baskar_ka@yahoo.com, sakthihit@gmail.com
* Corresponding author: rajsugan20@gmail.com

DOI: 10.1201/9781003635178-19

impoverished nations are heavily dependent on agriculture. India's economy heavily relies on the agricultural industry. The agriculture sector of the Indian economy is responsible for providing food and other essential commodities. Hence, any damage to the crop leads to a significant decrease in production, exerting a substantial impact on the nation's advancement. Plant leaves exhibit a high degree of fragility. Plant diseases frequently manifest as leaf browning or wilting. The dimensions and calibre of a crop are contingent upon the well-being of the plants responsible for its production. Hence, the timely identification of plant diseases on foliage is of utmost importance [4]. The expeditious containment and prevention of disease transmission within a farm yield superior outcomes. For example, by analyzing leaf type and condition, farmers can identify plant diseases. The successful implementation of this technique requires a level of talent and effort beyond the average. However, it would not be feasible on farms of such immense size. Modifications to symbols, hues, and other visual indicators are carefully examined for indications of pollution. A variety of image processing techniques are available for the purpose of identifying sick leaves [5]. Several image processing methods can be employed to spot diseased leaves. The picture illustrates various instances where image processing is employed to spot the leaf diseases. The process of identifying plant diseases is illustrated in Figure 14.1.

The aforementioned stages are elucidated as:

The act of capturing a photograph of any object in the physical world is referred to as "image acquisition." In this situation, the phrase is employed to denote the process of recovering a lost photograph. The initial stage in recovering the picture involves the hardware [7]. The aforementioned efforts are undertaken in anticipation of this possibly pivotal source in subsequent periods. The commencement of the imaging system's workflow is contingent upon the initial capture of an image. The entirety of the collected video is authentic.

Pre-processing is a step in the image-processing pipeline that involves correcting any unwanted image distortions and improving specific image characteristics. During this phase, it is possible to modify the size and type of the image in real-time, use denoising techniques, translate the picture, improve the quality of the shot, and execute morphological operations [8].

The act of segmenting an image involves the division of the image into smaller, more easily controllable portions. The core of this technique revolves around the extraction of information and the recognition of objects. Through the utilisation of this methodology, the camera assesses the resultant image [17]. This methodology is employed for the purpose of identifying the boundaries and constituents of an image. To effectively annotate an entire image, we leverage adjacent pixels that possess a comparable label section. The images can be segmented using KMC, Otsu's technique, or thresholding approaches.

The process of feature extraction results in the creation of a "region of interest" (RoI). Consequently, we shall proceed to analyse the RoI and examine its qualities. This methodology involves the extraction of a set of attributes, referred to as features, from an image with the purpose of generating image-related data and

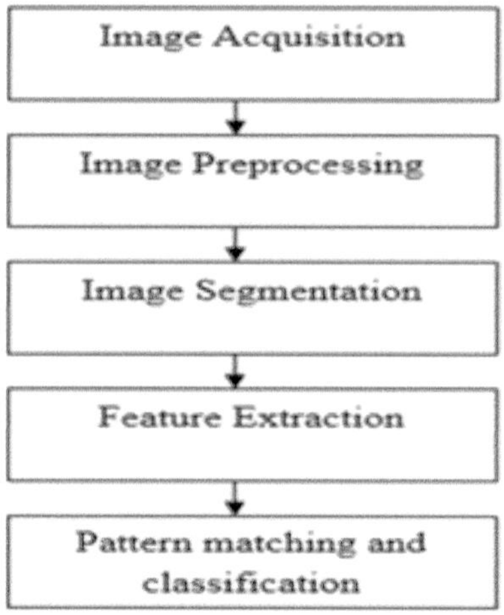

Figure 14.1 Various Stages in Plant Disease Recognition.

performing subsequent processing. A diverse array of indicators can be employed to detect plants that have undergone contamination. There are various techniques available for extracting attributes. Examples of feature extraction approaches include histogram-based methods [9] and graphical linear discriminant analysis (GLCM). It is employed for the statistical categorization of textures.

Pattern recognition classification is a method used to assess a provided sequence of symbols with the objective of preserving the fundamental characteristics of a pattern while facilitating its matching. The approach is predicated upon juxtaposing one's personal characteristics with those depicted in the photos provided within the repository. The patterns are categorised based on the types of deductions that can be drawn from them. Firstly, we will examine two separate approaches to classification: supervised and unsupervised learning algorithms. During the training process, humans are responsible for selecting the illustration pixels for the first-generation classifiers.

2. Literature Review

The study conducted by Abirami Devraj et al. examined the significance of agriculture in addressing the demands of an expanding population. Agriculture is considered one of the primary sources of income for more than 70% [11] of Asia's inhabitants. Nevertheless, the crops' quality was compromised by various additional illnesses. The agriculture industry could have remained unharmed with the accurate detection of the illness. The primary aim was to develop a software application capable of autonomously identifying and classifying diseases. The disease was progressively identified. Leaf pictures were employed to facilitate the identification of plant diseases. Hence, image processing techniques can be employed in agriculture to categorise plant diseases.

A study conducted by Meena Prakash et al. utilised image processing techniques [12] for the purpose of identifying diseases within plant leaves. The primary aim of this learning is to categorise leaf diseases using image analysis techniques. The system developed comprises four distinct stages, including preprocessing of the picture, segmentation of the image, feature extraction from the image, and classification of the image based on condition [18]. The photos of the leaves were segmented using the K-means algorithm for clustering in order to identify the precise locations of injury. Additional texture attributes were extracted using a Gray-Level Co-occurrence Matrix (GLCM) technique. A Support Vector Machine (SVM) was employed to classify the many kinds of leaves seen on plants.

According to Sharath DM et al., the recognition of plant illnesses is vital in agriculture due to the significant impact they have on output losses. The quality of farm goods was impacted by this phenomenon [13]. Conducting manual assessments of plant health and identifying signs of pest infestations proved to be an arduous and time-intensive endeavour. The correct diagnosis of plant diseases necessitates specific expertise. The aforementioned approach was also considerably time-intensive. Consequently, the field of image processing has gained significance in the diagnosis of plant diseases. The method had multiple phases. The conducted tests facilitated the monitoring of the plant's recuperation during its struggle against the sickness. The present study conducted an analysis of digital photographs of damaged plants in order to determine the optimal approach for the identification of plant illnesses.

In their study, Gurleen Kaur Sandhu et al. examined and deliberated on various techniques for employing photographs to detect plant diseases. In recent decades, numerous researchers have conducted investigations on these systems due to their potential in the identification of plant diseases [14]. Notable mechanisms included BPNN, SVM, KMC, and others. Algorithms successfully identified the health status of plant leaves. Nevertheless, the system had certain deficiencies. These weaknesses arose from the implementation of computerization in identification frameworks, including the use of external light and intricate photos obtained in challenging climatic conditions. The efficacy and efficiency of the current procedures have been demonstrated. Notwithstanding certain constraints, these methodologies have the potential to serve as the fundamental framework for a complete system aimed at the identification of plant diseases. Hence, additional investigation is necessary to enhance the existing corpus of knowledge in this field.

Sachin D. first delineated the latent primary disease finding in plant leaves to mitigate crop damage and enhance cultivated output [15]. The field of plant pathology was founded upon the examination of plants exhibiting unique traits. In order to achieve sustainability in agriculture, it is imperative to closely monitor the well-being of plants and actively detect any indications of disease. The task of visually observing plants for indications of disease required a significant amount of exertion. The success of this strategy relied on the crucial factors of competence and attentiveness in the study of plant diseases. This process had an equally long duration. Hence, the necessity for enhanced picture processing technology arose. The identification of plant diseases necessitates a series of procedures, encompassing picture acquisition, preprocessing, segmentation, feature extraction, and classification. This study elucidated the methodology employed for the identification of plant infections through the utilisation of leaf photos. Furthermore, this book delineated several methodologies for the detection of plant diseases.

In their study, Anand et al. (2016) developed a novel approach for the identification of diseases that are transmitted through plant leaves. The state-of-the-art approach effectively detected diseases [16]. The brinjal plant's leaves were analysed for disease by means of an amalgamation of image processing and artificial neural network (ANN) methods. Brindle leaf rot posed a significant problem. The production of brinjal had a significant decline due to this factor. The treatment of the brinjal plant was limited to its leaves, rather than encompassing the entire plant [19]. The brinjal leaves were shown to be the principal host in around 85–95% of cases. The images were split utilising the KMC methodology and subsequently categorised employing NN. The artificial neural network (ANN) framework effectively identified leaf illnesses in brinjal plants.

3. Research Methodology

The process of plant illness recognition involves utilising leaf images to identify the location of infection. This paper presents several methodologies to enhance the diagnostic capabilities of plant diseases. The diagram in Figure 14.1 illustrates the proposed framework for finding plant diseases.

Conditioned: The first step in the recognition of plant illnesses is the process of picture pre-processing. The Plant Village collection is preferred due to its reliability and comprehensive coverage of plant information and the diseases that affect them. The collection includes images depicting potato leaves in both healthy and blight-affected conditions. Prior to any further manipulation of the images, it is essential to convert them from RGB to grayscale.

Segmentation: Segregation is the subsequent step in the process of creating plant images for the purpose of item identification or data extraction. Segmentation enhances the manageability of image analysis by examining the boundaries of images and objects. K-Means Clustering (KMC) can be employed for object classification due to the unique properties exhibited by pixels with identical labels. To segment pictures, a larger number of k (three) is used after determining the necessary area based on the disease forecast from the input leaf.

Feature Extraction: The method of Feature Extraction entails the elimination of specific features from the acquired Region of Interest (ROI). Extracting properties such as colour, texture, and other characteristics can assist in identifying the afflicted sections of plants. The correlation on a saturation scale matrix is a statistical tool used to determine textural properties. Early indicators can be identified by retrieving properties such as contrast, IDM, entropy, and so on. Below is an elucidation of these attributes:

Contrast is used to assess the level of detail and distinction in an image.

$$Contrast = \sum_{n=0}^{G-1} n^2 \left\{ \sum_{i=1}^{G} \sum_{j=1}^{G} P(i,j) \right\} . |i-j| \tag{1}$$

Subsidiary inputs from P(i,j) orthogonal to the sloping (ij) are a main function of this module.

Homogeneity: It is a degree of how close the GLCM's component dispersal is to the crosswise.

$$Homogeneity = \sum_i \sum_j \frac{P_d \lfloor i,j \rfloor}{1+|i-j|} \tag{2}$$

Local Homogeneity, IDM:

$$IDM = \sum_{i=0}^{G-1} \sum_{j=0}^{G-1} \frac{1}{1+(i-j)^2} P(i,j) \tag{3}$$

Less weight is given to the Inhomogeneous regions (ii) due to the Inverse Difference Moment (IDM), due to the weight factor (1+(i–j)2) (–1) IDM. There is a lack of homogeneity in the picture, which will have a lower IDM value, while the pictures that have greater sense will have a higher value of IDM.

The quantity of information that is present in the picture is quantified by entropy, by which the intensity distribution and unpredictability is been measured.

$$Entropy = \sum_{i=0}^{G-1} \sum_{j=0}^{G-1} P(i,j) \times \log(P(i,j)) \tag{4}$$

Correlation is a characteristic that is used to regulate the dependency of the linear gray-level with the pixels in a particular position in relation to one another.

$$Correlation = \sum_{i=0}^{G-1} \sum_{j=0}^{G-1} P \frac{(i \times j) \times P(i,j) - \{\mu_x \times \mu_y\}}{\sigma_x \times \sigma_y} \tag{5}$$

Sum of Squares, Variance:

$$Variance = \sum_{i=0}^{G-1} \sum_{j=0}^{G-1} (i-\mu)^2 P(i,j) \tag{6}$$

The need for allocation of larger weights to the components is underlined by this feature, which helps to distinguish based on the normal value of P(i,j).

Sorting Information: The development of a technique for the identification of plant diseases is an essential step in the progression of sorting information. A training set and a test set are created by dividing the entire dataset. The introduction section of the document accounts for approximately 60% of the information, and the balance 40% is allocated for the ending disease categorization. The determination is predicated upon the outcomes of a statistical classifier.

The enhanced accuracy of this classifier can be attributed to the integration of decision trees, random forests, and Gradient boosting techniques. The utilisation of a decision tree can be considered a viable approach for decomposing the entire set of options into smaller, more controllable groups.

The predetermined decision-making standards are utilised to differentiate each subcategory from the remaining subgroups. Decision trees are a great tool when analysing data that includes hierarchical and fixed-target variables. In the DT algorithm, both the assortment criteria and the stumbling rule play key roles.

Various metrics such as the Gini index, data entropy, and misclassification error can be employed to refine the range of potential outcomes. In situations when the objective is predetermined, a frequently employed criterion is to reduce the square of the error. Initially, DT analyzes the complete dataset of training data in order to identify each of the prognostic elements in every possible gap. This will continue indefinitely until the restrictions on withholding are enforced. Since the DT is non-parametric and does not assume a positive distribution, it does not require a functional method for its predictors.

Due to the potential impact of even little alterations in the training data on the structure of the tree, decision trees are inherently flawed. There exists a notion suggesting that the amalgamation of outcomes from many trees could potentially yield advantages in terms of volatility and the identification of previously overlooked signals within the dataset. RF is categorised as an ensemble framework because it combines multiple performance-enhancing designs. In contrast to a solitary tree, the several trees in RF exhibit a greater degree of disorganisation as they engage in competition to assert their dominance. A multitude of causes exerted an effect on the evolution of these trees. Linking methods or branches of a tree. Figure 14.2 illustrates the construction and implementation of the proposed paradigm.

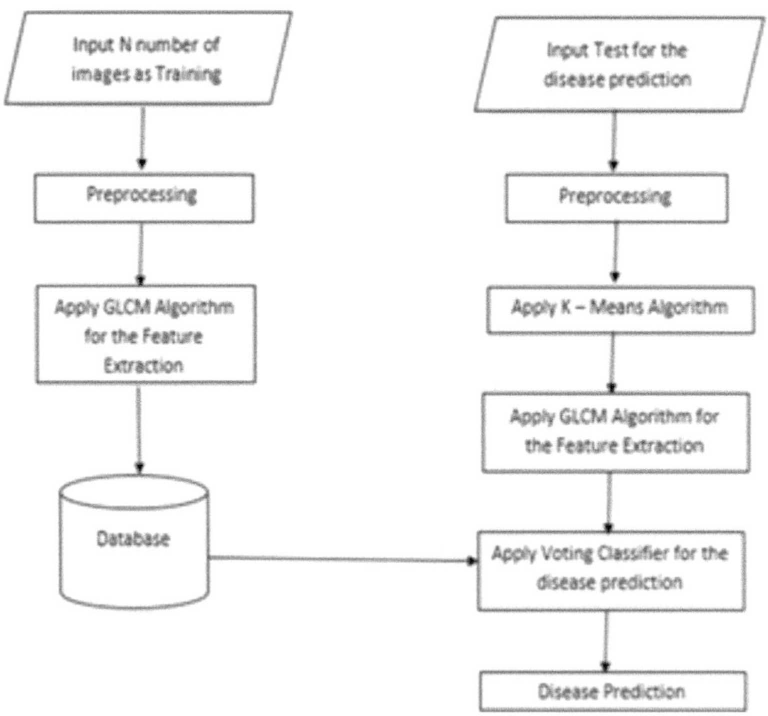

Figure 14.2 Planned Model.

4. Result and Discussion

The employed methodology leverages the openly available Plant Village dataset, which has a wealth of information pertaining to various plant diseases. Each photo in this series is accompanied by a concise explanation of the related medical concern. During the developing phase, retinal representations of potato leaves are employed. There are three distinct categories in which images can be classified: those portraying healthy conditions, those portraying the initial phases of blight, and those portraying the advanced stages of the illness. Various measurements are employed to evaluate performance, such as memory, precision, and accuracy, all of which are necessary. Below, we will delve into these defence measures in further detail:

A. The percentage of appropriately recognized trials is related to the total number of trials in a specific programme, which is what is intended by "accuracy." The rate of the constraint may be articulated mathematically as:

$$A_i = \frac{t}{n} \cdot 100 \tag{7}$$

To compute the quantity of right classifications, t, it is essential to recognize how many trials are used (n).

B. *Precision*: This metric demonstrates how many accurate positive results there were, relative to the total number of positive results.

C. *Recall*: This metric is found by dividing the TPs by the total positive cases.

D. *Voting Classifier*: A voting classifier is an algorithm for learning that predicts an output (class) based on the class that has the best chance of becoming the output. It acquires experience by training on a set of several models. It averages the output class predictions from each classifier that is fed into the voting classifier in order to predict the output class with the biggest majority of votes.

Instead of creating distinct, specialised models and assessing each one's correctness, the idea is to create a single model that learns from several models and forecasts output based on the combined majority of votes for each output class.

The accuracy, precision, and recall of SVM, KNN, MLP, Random Forest, and the proposed model are presented in Table 14.1. The efficiency of the recommended model is related to that of the SVM, KNN, MLP, Random Forest models, and the proposed model in Figure 14.3. The anticipated model outdoes state-of-the-art models by coarsely 5% in terms of accuracy, precision, and recall is shown in Figure 14.3.

Table 14.1 Comparison of Different Classifiers' Performance.

Classifiers	Accuracy	Precision	Recall
SVM	76.772%	61%	77%
KNN	63.962%	58%	62%
MLP	67.661%	57%	68%
Random Forest	82.484%	82%	82%
Proposed Model	84.183%	84%	84%

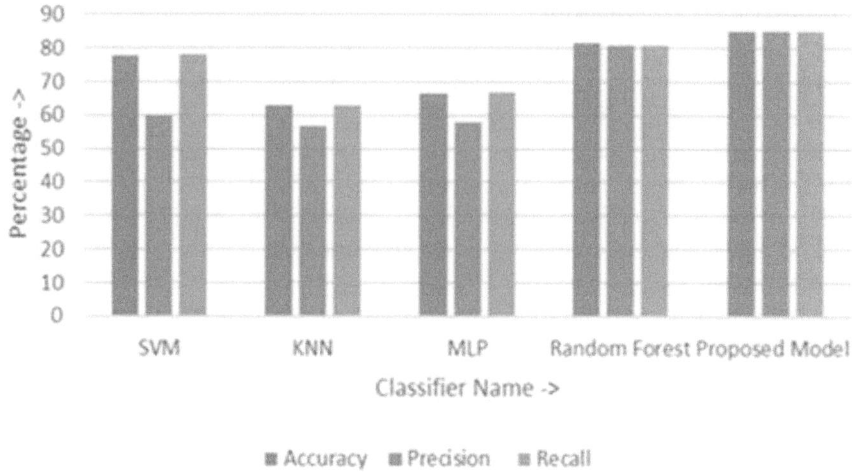

Figure 14.3 Evaluation of different classifiers' performance.

5. Conclusion

The main aim of this investigation is to generate and authorize an end-to-end system for the identification of plant diseases. A comprehensive system has four distinct phases, including pre-processing of the image, attribute extraction, image segmentation, and sickness categorization. In the past, microscopy was essential for the identification of plant diseases. The approach was once highly effective, but it became less popular due to the difficulties associated with microscopic leaf analysis. The K-Means Clustering (KMC) technique is employed for image feature extraction, while the Gray-Level Co-occurrence Matrix (GLCM) strategy is utilised for segmentation. The Support Vector Machine (SVM) technique will be replaced by the new technique. The experimental methodology includes integrating a voting classification system. By employing three metrics, we find that the new approach achieves 10% higher accuracy and a lower False Positive Rate (FPR) than the conventional strategy.

References

[1] Rekha Balodi, Sunaina Bisht, Abhijeet Ghatak and Rao, K.H. (2017). Plant disease diagnosis: technological advancements and challenges. Indian Phytopath. 70(3): 275–281.

[2] Monishanker Halder, Ananya Sarkar and Habibullah Bahar. (2019). Plant disease detection by image processing: a literature review. SDRP Journal of Food Science & Technology.

[3] Surender Kumar and Rupinder Kaur. (2015). Plant disease detection using image processing: a review. International Journal of Computer Applications 124(16): 0975–8887, August 2015.

[4] Sabah Bashir and Navdeep Sharma. (2012). Remote area plant disease detection using image processing. IOSR Journal of Electronics and Communication Engineering (IOSRJECE) 2(Issue 6) (Sep-Oct 2012): 31–34.

[5] Rajneet Kaur and Miss. Manjeet Kaur. (2017). A brief review on plant disease detection using in image processing. International Journal of Computer Science and Mobile Computing 6(Issue. 2) February 2017: 101–106.

[6] Sushil R. Kamlapurkar. (2016). Detection of plant leaf disease using image processing approach. International Journal of Scientific and Research Publications 6(Issue 2) February 2016.
[7] Santhosh Kumar, S. and Raghavendra, B.K. (2019). Diseases detection of various plant leaf using image processing techniques: a review. 2019, 5th International Conference on Advanced Computing & Communication Systems (ICACCS).
[8] Saradhambal, G., Dhivya. R., Latha, S. and Rajesh, R. (2018). Plant disease detection and its solution using image classification. International Conference on Advanced Computing & Communication Systems (ICACCS) 119(14): 879–884.
[9] Sujatha, R., Sravan Kumar, Y. and Garine Uma Akhil. (2017). Leaf disease detection using image processing. 2017, Journal of Chemical and Pharmaceutical Sciences 10(Issue 1).
[10] Tejal Deshpande, Sharmila Sengupta and Raghuvanshi, K.S. (2014). Grading & identification of disease in pomegranate leaf and fruit. IJCSIT 5(3): 4638–4645.
[11] Devaraj, A., Rathan, K., Jaahnavi, S. and Indira, K. (2019). Identification of plant disease using image processing technique. 2019 International Conference on Communication and Signal Processing (ICCSP), pp. 0749–0753.
[12] Prakash, R.M., Saraswathy, G.P., Ramalakshmi, G., Mangaleswari, K.H. and Kaviya, T. (2017). Detection of leaf diseases and classification using digital image processing. 2017 International Conference on Innovations in Information, Embedded and Communication Systems (ICIIECS), pp. 1–4.
[13] D.M., S., Akhilesh, S.A., Kumar, R.M.G. and C., P. (2019). Image based plant disease detection in pomegranate plant for bacterial blight. 2019 International Conference on Communication and Signal Processing (ICCSP), pp. 0645–0649.
[14] Sandhu, G.K. and Kaur, R. (2019). Plant disease detection techniques: a review. 2019 International Conference on Automation, Computational and Technology Management (ICACTM), pp. 34–38.
[15] Khirade, S.D. and Patil, A.B. (2015). Plant disease detection using image processing. 2015 International Conference on Computing Communication Control and Automation, pp. 768–771.
[16] Anand, R., Veni, S. and Aravinth, J. (2016). An application of image processing techniques for detection of diseases on brinjal leaves using k-means clustering method. 2016 International Conference on Recent Trends in Information Technology (ICRTIT), pp. 1–6.
[17] Thirumeninathan, V., Vijayalakshmi, S. and Palathara, S.T. (2024). Enhancing banana cultivation: disease identification through CNN and SVM analysis for optimal plant health. 2024 International Conference on Trends in Quantum Computing and Emerging Business Technologies, Pune, India, pp. 1–6, doi: 10.1109/TQCEBT59414.2024.10545264.
[18] Lekha, J. and Vijayalakshmi, S. (2024). Enhanced weed detection in sustainable agriculture: a you only look Once v7 and internet of things sensor approach for maximizing crop quality. Engineering Proceedings, Open Source Preview 82(1): 100.
[19] Thomas, S. and Vijaylakshmi, S. (2022). Image recognition, recusion cellular classification using different techniques and detecting microscopic deformities. 2022 2nd International Conference on Advanced Computing and Innovative Technologies in Engineering (ICACITE), Greater Noida, India, pp. 1053–1055, doi: 10.1109/ICACITE53722.2022.9823758.

Blockchain and System Performance

Chapter 15

Optimising Permissioned Blockchain Performance With Topology-Aware Raft Consensus

Kiran Kumar Kondru[1,*] and *Saranya Rajiakodi*[2]

1. Introduction

The use of blockchain technology has been on the rise, especially within the realm of large-scale contemporary corporations. Public as well as permissioned/consortium blockchains have found application across diverse sectors beyond the scope of digital currencies. Corporations exhibit a stronger inclination towards consortium blockchains like Hyperledger Fabric, which provides functionalities such as smart contracts in collaboration with a range of business partners.

With the increase in adoption, these drawbacks are becoming more visible, especially regarding scalability. As the quantity of participating nodes rises, the throughput of the transactions continues to decline drastically, especially for permissioned blockchains.

Permissioned/consortium blockchains are much faster than public blockchains such as Bitcoin and Ethereum, as they depend on proof-based consensus like proof-of-work, proof-of-stake, etc. As these public blockchains operate in a Byzantine environment where there are nodes with malicious intent, extra cryptographic security measures are included in the design itself so that the integrity of the transaction blocks is kept intact. This added proof-based consensus mechanism adds complexity

[1] PhD Research Scholar, Department of Computer Science, Central University of Tamil Nadu, India.
[2] Assistant Professor, Department of Computer Science, Central University of Tamil Nadu, India.
* Corresponding author: kirankondru.tech@gmail.com

DOI: 10.1201/9781003635178-21

and time- and energy-consuming techniques to negate malicious attacks, which eventually slows down the throughput of the blockchain network. According to some calculations, Bitcoin handles up to 4.6 TPS, whereas Visa handles 1,736 TPS. By delegating security measures outside of blockchain consensus, permissioned blockchains have gained an advantage over public blockchains. This is mainly because of the use of traditional distributed consensus, which is usually deployed in distributed databases and datastores. Traditional distributed consensus systems such as Paxos and Raft offer greater throughput, leading to much faster block creation and commit. Hyperledger Fabric, one of the first and most well-known permissioned blockchains, uses a raft implementation for its ordering service, which has been thoroughly tested.

The raft consensus algorithm (1) is a robust leader-centric protocol in which the leader receives commands from clients, transmits them, and ensures that followers commit to the transactions. Raft was specifically crafted to enhance comprehensibility and implementation, in contrast to the intricacy of the Paxos consensus, despite its mathematical validation. Due to its user-friendly nature, the raft protocol has been more swiftly embraced in both industry and academia compared to Paxos, sparking significant research interest (2,3). The adoption of raft consensus in Hyperledger Fabric, as opposed to Paxos, serves as a testament to its accessibility. Formal verification of raft was carried out using TLA+(4) and LNT. Efforts have been undertaken to integrate forensic (5) techniques for a deeper insight into the protocol.

The broader utilization of permissioned blockchains like Hyperledger Fabric has led to the observation of performance deterioration when numerous participating nodes are involved in the blockchain network. This issue of scalability has long existed, but only recently started to receive attention. Conventional distributed consensus algorithms are tailored for databases, particularly distributed databases or datastores (6), and typically do not require many nodes to cater to clients. Nevertheless, with the emergence of blockchain technology, the potential to increase the number of nodes within the network has risen notably. Once a specific threshold is exceeded, the traditional consensus algorithm degrades rapidly. In the subsequent sections, we deliberate attempts to increase throughput and scalability.

2. Related Works

Several efforts have been undertaken to alleviate the performance decline in conventional consensus protocols, particularly in the context of Paxos extensions. The authors achieved adequate outcomes. The enhancements primarily focus on refining the primary consensus protocols. This article delves into various consensus mechanisms and their endeavors to address scalability challenges.

i) **Paxos:** Paxos (7) is widely used, and many variants already exist. In each round, there is a new leader.

ii) **Fast Paxos:** Fast Paxos (8) aimed to improve Paxos, reducing the number of rounds required to pass messages, achieving higher speed.

iii) **Generalized Paxos:** Generalized Paxos(9) allows nodes to execute out-of-order requests to improve performance.

iv) **FGCC:** FGCC (10) builds upon Generalized Paxos and reduces the message complexity to resolve conflicts.

v) **Ring Paxos** (11) and **Multi-Ring Paxos** (12)**:** The acceptors were organized along a ring. However, as the number of nodes in the ring increases, the latency increases.

vi) **EPaxos, Mencius** (13), and **S-Paxos** (14)**:** They are more decentralized than the aforementioned protocols, but they also broadcast new messages in each round, overwhelming the underlying network.

3. Raft Consensus Algorithm

The Raft consensus algorithm has been intricately crafted with the intention of enhancing comprehension, particularly when compared to the renowned Paxos consensus. It is formulated as a robust leader-centric consensus protocol, in which the leader plays an active role in all decisions within the Raft cluster, while the remaining servers in the group function as followers that replicate the directives issued by the leader in the specified sequence. When a leader fails, followers of the group elect a new leader with a new term number. This leader-based scheme of controlling the group is easier to think about and implement than Paxos, which is mathematically proven, but developers find it harder to develop and deploy.

Although the Raft protocol came out only in 2015, the number of citations was comparable to that of the more recognized Paxos consensus. Even GitHub. com downloads Raft implementations, and adoption of the Raft algorithm in modern distributed databases has risen significantly. Raft is also taught in numerous universities, as it is easier for students to understand, learn, and implement.

While Raft's performance can be likened to that of Paxos, it is noteworthy that scalability challenges persist. The performance degradation accelerates notably as additional nodes are integrated into the cluster. The leader is responsible for receiving client requests, organizing these commands in sequence, and disseminating them to the follower nodes. Subsequently, upon the acceptance and replication of commands by the majority of nodes, the leader transmits a response back to the client.

3.1 Working of the Raft Algorithm

The Raft algorithm constitutes a robust leader-centric consensus mechanism, albeit not centralized, since any supporter possesses the potential to ascend to the leadership role within the cluster. Upon the initiation of operations within the cluster, all nodes within the collective commence their roles as followers. Subsequently, one of the supporters awakens due to a random timer expiration and recognizes the absence of a leader within the group, prompting a transition to a candidate status. This action results in an increment of the term number, accompanied by the transmission of vote requests to other followers. The followers, in turn, respond with affirmative votes to the candidate in a sequential manner. Following the acquisition of a majority of votes ($n/2 + 1$), the candidate transitions into a leader, commencing the dissemination of broadcast heartbeat messages to inform all supporters of the leader's identity and the current term.

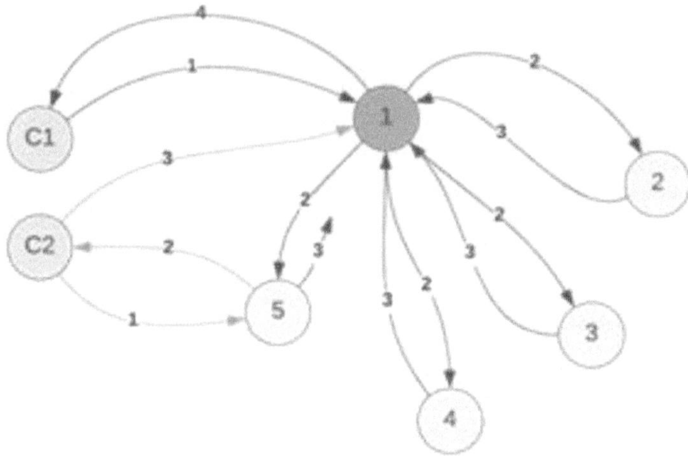

Figure 15.1 Functioning of a Raft cluster after a client's request.

The Lead is responsible for managing all client instructions, organizing them in a specific order, and disseminating them to the followers. Subsequently, the Leader anticipates feedback from the majority of supporters and transmits an authorisation message back to the client. These actions encapsulate the fundamental process of the Raft consensus mechanism. The diagram depicted below mirrors this process: depicted in Figure 15.1 are clients C1 and C2. Client C1 solicits leader "1" to establish a value, a request that "1" propagates to nodes 2, 3, 4, and 5. While nodes 2, 3, and 4 acknowledge "1" affirmatively, the acknowledgement from follower "5" is not received. Nevertheless, this setback does not impede the leader's progress, since it has garnered responses from the majority of followers (2, 3, and 4) excluding itself. Consequently, the leader deems the request as officially approved by the cluster, and in the final stage (4), it communicates with client C1.

In Figure 15.1, client C2 tries to make a read request to node 5, which is a follower, but it responds by saying go to leader "1." Then, C2 contacts leader "1." This provides the context of how the Raft algorithm works.

3.2 Raft Availability

Raft has strong consistency guarantees in that the whole cluster maintains the same truth. From the client's point of view, the cluster can serve almost always. Even if some nodes fail or the network is partitioned, the cluster still handles the situation gracefully. Even when a leader node fails, one of the followers wakes up and starts the election. If it receives enough votes from others, it becomes the leader and continues to serve clients. In the following simulation, we attempted to fail the leader node 5% of the time; however, the entire cluster became available and stable.

Figure 15.2 shows the histogram. Even though the leader fails 5% of the time, when the leader is elected, their stability in terms of the time they spent being the leader is shown. The histogram shows that leaders stayed as the head of the cluster for more than 1,000–1,900 units of time. And this histogram also shows that the term

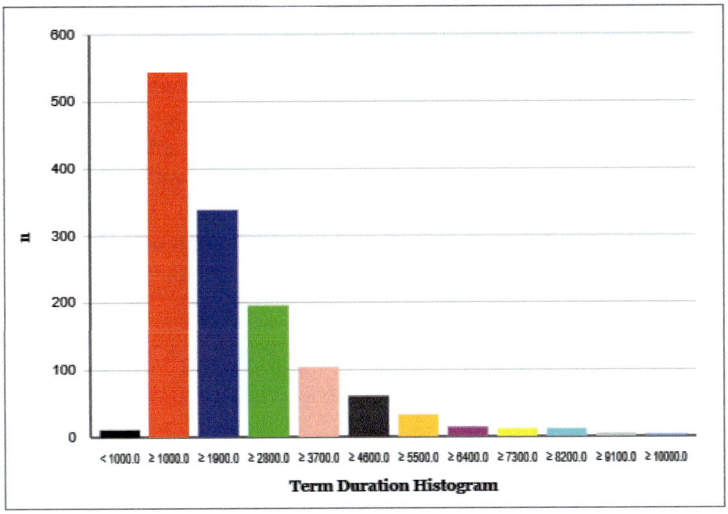

Figure 15.2 The amount of time the whole cluster is alive.

time is the least, with less than 1,000 units of time. This can show that the cluster is alive most of the time, even with leader failures.

The above figures show the time-duration histogram of the total raft cluster. Even with a leader failing 5% of the time, the entire cluster becomes operational, and most of the time, it is available. Be it a 5-node cluster or a 13-node cluster, a new leader emerges when a leader fails. As we have shown through simulation, the raft's availability remains good, even with an increased quantity of nodes in the cluster. But, throughput decreased as the number of nodes increased.

3.3 Issues With Throughput and Scalability

The following are the issues with Raft that have already been recognized.

- There exists a singular node, known as the leader, which attends to all client requests, arranges them in the sequence of commands received, disseminates the commands to supporters for replication, collects the majority consensus from followers to commit to their log effectively, and subsequently notifies the client of the effective fulfillment of the request. Consequently, this results in the leader operating at a slower pace and establishing a hypothetical threshold on the number of requests it can manage within a designated timeframe.
- As the quantity of nodes within the cluster expands, there is a corresponding escalation in the frequency of broadcasts, consequently amplifying the leader's obligation to await responses from a substantial portion of the nodes. This phenomenon establishes a theoretical threshold regarding the extent to which the cluster size can expand before observing a notable decline in the cluster's operational efficiency.

- After the replication process is completed, the leader must respond to the client, and no follower is allowed to do so. For writing requests, it is imperative that the leader ensures that the commands are properly replicated, and the read requests can be accepted by any follower and respond. But there is a caveat. Given a scenario where a client C1 request the value of a variable "x" and at the same time, another client C2 requests to change of the value of "x" from to "1" to "2," there is a change that C1 would receive the value for "1" x for "x" instead of the updated value of "2." This is called a stale read, wherein not all clients obtain the updated value of "x." Raft avoids this by doing so through the leader. However, this also limits the throughput.

In order to address these concerns, a pioneering approach is suggested to enhance the efficiency and capacity of the Raft cluster.

4. System Model

As delineated in the preceding section, the challenges pertaining to throughput and scalability are methodically addressed in this discussion, alongside the proposition of a remedy aimed at significantly mitigating the decline in performance with the expansion of nodes and enhancement of throughput.

4.1 Assumptions in Raft Consensus

The raft Consensus makes certain assumptions about the environment in which it is going to work. The following assumptions are made (15):

i) *Unreliable network*: The network conditions are unreliable, including delays, network partitions, packet loss, re-ordering, duplication, etc.
ii) *Non-Byzantine Environment*: Raft assumes there are no malicious nodes.
iii) *Static Configuration*: Nodes are statically configured, and each node knows the address of every other node.
iv) *Clients*: Clients only communicate through the leader
v) *Complete Network*: Raft assumes a complete graph, wherein all the nodes are connected to all other nodes

4.2 Proposal Description

As stated in the previous section, how the leader alone is responsible for accepting, sequencing, broadcasting, and finally responding to the client, we propose to delegate some of the broadcasting responsibilities to the rest of the follower nodes. As stated above, Raft assumes a complete graph in which every node is linked to every other node, which is not feasible in reality. Raft is designed for a closed-connected cluster of servers in datacenters. However, the actual network topology differs from that assumed by Raft. Data centers usually follow a close-like topology, such as fat-tree, VL2, or lead/spine. This underlying network forwards packets to the nodes in the cluster, although the overlay network might be in the form of a complete graph.

For each command issued by a client, there is a replication process involving dissemination from the leader and subsequent acknowledgment. The concept of virtual dissemination operates under the assumption that each node has a direct link to the leader; nonetheless, in practice, the routing procedures are managed by the network infrastructure, including switches. At times, certain switches may be strained while others remain underutilized. The Raft consensus algorithm, akin to Paxos, is built on the premise of a one-to-one interconnection among all network nodes. Given the absence of a tangible physical network configuration, this assumption serves as a beneficial abstraction.

In our proposal, we refrain from assuming a particular network topology, as there exists a multitude of possibilities. Instead, we critique the impractical one-to-one complete graph strategy. Our approach involves the transmission of tree-like messages among the servers within the cluster to address this issue. A similar design principle is also observed in Layered Paxos (16).

i) *Delegating the responsibility of broadcasting*: Rather than broadcasting to all nodes within the cluster, the leader opts to broadcast solely to a specific subset of nodes in the network, which subsequently disseminates the broadcast to another select subset of nodes. The propagation process necessitates only two tiers if the chosen nodes are appropriately selected. Further elucidation on this topic will be provided below.
ii) In the original raft algorithm, exclusively the leader is permitted to respond to clients after a command being added to the cluster's log as shown in Figure 15.3. This stipulation has been loosened to allow any follower node linked to a client to handle a read request. It has been highlighted in preceding sections that there

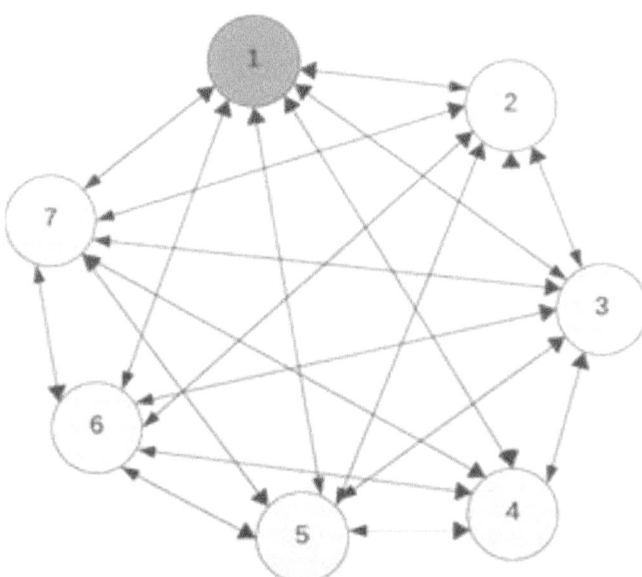

Figure 15.3 7-node Raft cluster, with one-to-one overlay network.

exists a risk of outdated reads. To address this concern, a delayed response is recommended, given the 2-level broadcast process from the leader; waiting for both levels to finalize would prevent encountering a stale read.

4.3 Two-Level Broadcast Propagation

In the initial section of the suggestion, the assignment of disseminating the repetition command (Append Entries) to specific nodes that are pre-determined within the cluster is undertaken. Subsequently, these designated nodes propagate the broadcast to the remaining nodes within the cluster. The acknowledgment to this Append Entries command is transmitted back to the leader, who subsequently responds to the client. This process is visually depicted in Figure 15.4, featuring a Raft cluster consisting of seven nodes.

Clients C1 transmit a set(x,2) message to assign the value of "x" as "2" within the raft cluster. Leader 1, rather than disseminating the message to all six remaining supporter nodes, directs it solely to two other nodes, identified as 7 and 2. Node 7 subsequently communicates with nodes 6 and 5, relaying their responses back to leader "1." This process of multilevel broadcast message propagation may appear to involve additional rounds; however, it enables leader "1" to accommodate further clients' requests and repeat the process. Considering the configuration of the underlying physical network, this multilevel propagation alleviates the burden on certain overwhelmed switches and distributes the broadcast evenly throughout all networks.

As the magnitude of the network expands, the 2-tier dissemination can be upheld by simply augmenting the number of nodes that the leader engages with. When the count of nodes is either five or seven, appointing two proxy nodes proves to be

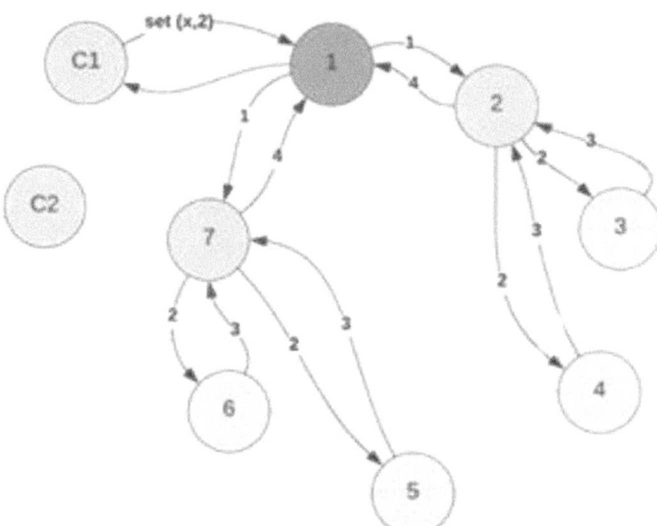

Figure 15.4 Proposed two-level replication broadcast.

adequate. In cases where the network consists of 9, 11, or 13 nodes, dissemination to three nodes becomes imperative. This assignment was executed to prevent any alteration in the security aspect of the Raft protocol. Consequently, each server is assigned the duty for a specific quantity of nodes denoted as "m." In scenarios where there are "n" nodes, the selection of "m" should be such that it satisfies the ensuing equation:

$$Log_{m}(n-1) \leq 2. \quad (1)$$

4.4 Delayed Local Read Request

Stale reads pose a significant challenge to data consistency. The Raft consensus algorithm is known for its stringent enforcement of consistency, ensuring that stale reads are effectively prohibited. This enforcement mechanism is carried out by the designated Raft leader, who diligently handles all read requests from clients. In practice, data systems often experience a higher volume of read requests compared to write operations, with many systems being specifically optimized to cater to read-intensive workloads. Our proposition revolves around enhancing the overall throughput of a Raft cluster by distributing the responsibility of handling read requests to follower nodes that are directly connected to clients.

We solved the problem of "stale" reads by delaying the read request to a certain limit. We take the 2-level broadcast into consideration and time it accordingly to the response to a read request. By this, though read requests delay the clients' requests a little, they are at least directly connected to the follower with the latest values in its data store. This reduces the load on the server by removing the extra sequencing (step) required by the leader. In (17), part of the compartmentalization is the replica to forward the read request.

4.5 Log Replication Algorithm

The following section describes the algorithm used for replicating the log. We show the changed algorithm for log replication here

PROXY LEADER:

- On receiving a delegate message from the Leader,
  o append entry to local log
- If last *indexLog* >= *indexNext* for FOLLOWER,
  o send ***AE*** with the *log* entries starting at *nextIndex*
    • If **success**: update *nextIndex* and *matchIndex* for a FOLLOWER
  o If ***AE*** **fails** because of log inconsistency:
    •Decrement nextindex and retry
- If there exists an **M** such that **M**> *indexCommit*,
  o a majority of *matchIndex*[i] >= **M** and *log[M].term* == *currentTerm*
    •Set *indexCommit* = **M**
- Respond with the result to the LEADER

In the algorithm described above, we delegated the duties of the leader to a proxy leader and made it responsible for checking with supporters to repeat the latest entry(s) from the leader. The above algorithm is kept simple to make it more understandable, in line with the original raft paper (1). The reader may refer to the complete algorithm mentioned in this paper because we are not reproducing the entire algorithm here.

4.6 Simulation

We implemented the Raft consensus algorithm in Java, coded a network simulator (in Java), and tested the protocol. First, we created a benchmark test using 25 Raft nodes. Initially, when the simulation starts, a leader election occurs, and log replication starts. We also coded for the 2nd-level leader (a proxy) who propagated the commands for the leader. We tried to simulate a hub-and-spoke model with the above simulations.

Our simulation is run on an Intel Core i3-4030 CPU with 8GB of memory running Windows 10 Operating System and JDK18. We used basic and open-source libraries, including java.io and java.net. We kept the setup simple so as to be easily reproducible, and no proprietary software is used here. However, we also acknowledge that the results might slightly vary depending on the configuration of the system used. The results are analysed and discussed in the next section.

5. Results and Analysis

Several overlay network topologies, such as Hub-and-Spoke, Partial Mesh, Lead-Spine topology, multi-tier architecture, Full Mesh, and End-to-End overlays, are at our disposal. In this instance, the Hub-and-Spoke topology was selected to demonstrate the operational dynamics of the raft replication process. Due to the assumption of a full mesh-like topology by the raft, the dissemination of information and corresponding responses appear as a unified occurrence, although, in reality, multiple network devices manage the data. While some routers may experience excessive load, others may remain predominantly inactive.

To get an idea of throughput decreases and latency increases, we ran a simulation of a raft cluster from 25 nodes to 50 nodes. The result is as depicted below in Figure 15.5.

We started testing from 25 nodes with increments of 5 nodes for each simulation run. We see a marked decrease in throughput by the time the number of nodes reaches 50. We also see almost a doubling of latency from 25 nodes to 50 nodes. This necessitates the need for improving throughput and latency to increase performance and make the cluster more scalable.

The following graphs illustrate both the benchmark results and the 2-level propagation results for the average throughput observed. We represented it with both a violin graph and a box plot, as the results included the results of multiple simulation runs.

The violin plot in Figure 15.6 and the box plot in Figure 15.7 both show the results of our simulation. We ran a simulation with 25 nodes. In the initial benchmark runs it was found that there is a significant increase in the throughput. We tested

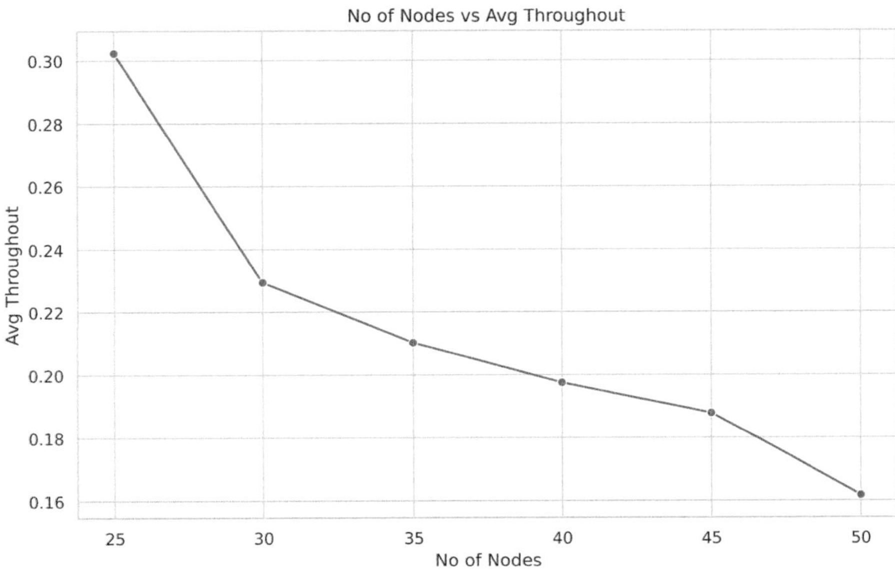

Figure 15.5 Number of Nodes vs Average Throughput.

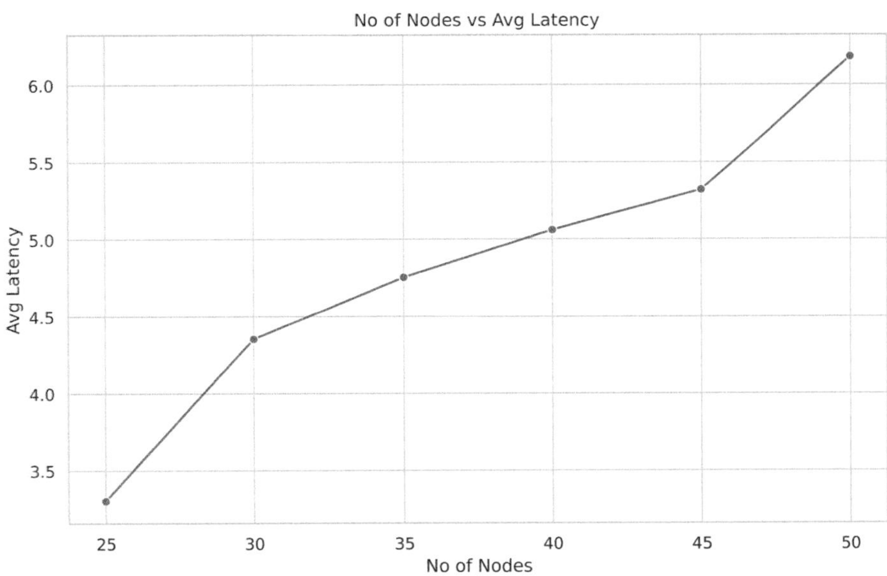

Figure 15.6 No of nodes vs average latency.

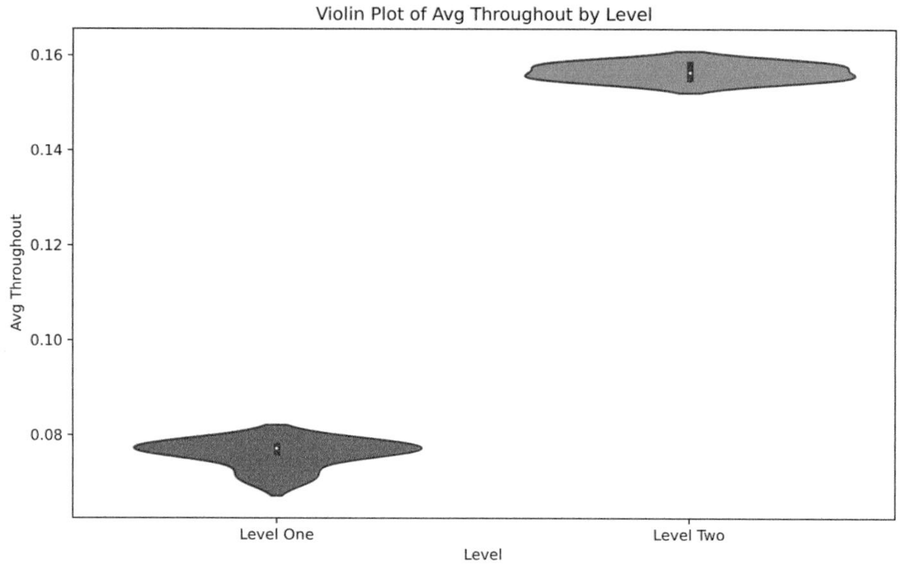

Figure 15.7 Violin plot of average throughput with 25 nodes - comparison.

with a single leader doing all the log replication and with two-level log propagation delegated to a proxy leader. The leader delegates the client's commands to the second (proxy) leader, who further communicates with the clients and replies to the leader. The leader, in turn, communicates this to the client. The table below shows the results for both levels.

Table 15.1 The table shows the results of 3 simulation runs for level one and 3 for level two.

Level	Duration	Average Throughput	Average Latency
Level One	42,059	0.078082	12.805
Level One	36,020	0.0763184	13.101
Level One	43,405	0.077196	12.949
Level Two	9,928	0.1546312	6.461
Level Two	6,433	0.1564455	6.388
Level Two	7,164	0.15842839	6.304

We also verified the latency with a 25-node cluster and found that a 2-level raft design reduced the latency in general, which is represented in the following box plot and violin graphs. As can be seen from the graph, average throughput significantly increased, while also marked decrease in latency as shown in Figure 15.8.

We also present an event-based series of diagrams that show the leader receiving client commands, broadcasting them, and replicating them through the cluster. This

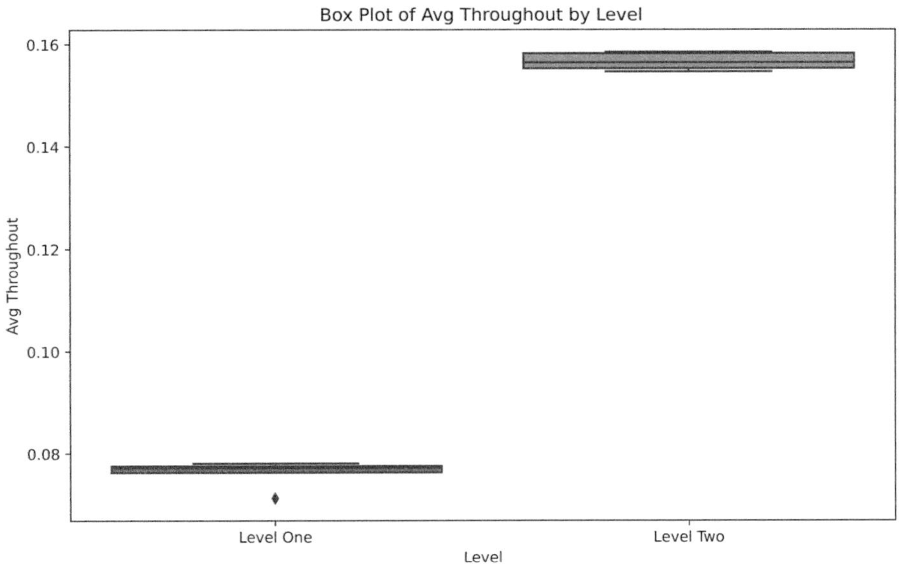

Figure 15.8 Box plot of average throughput comparison.

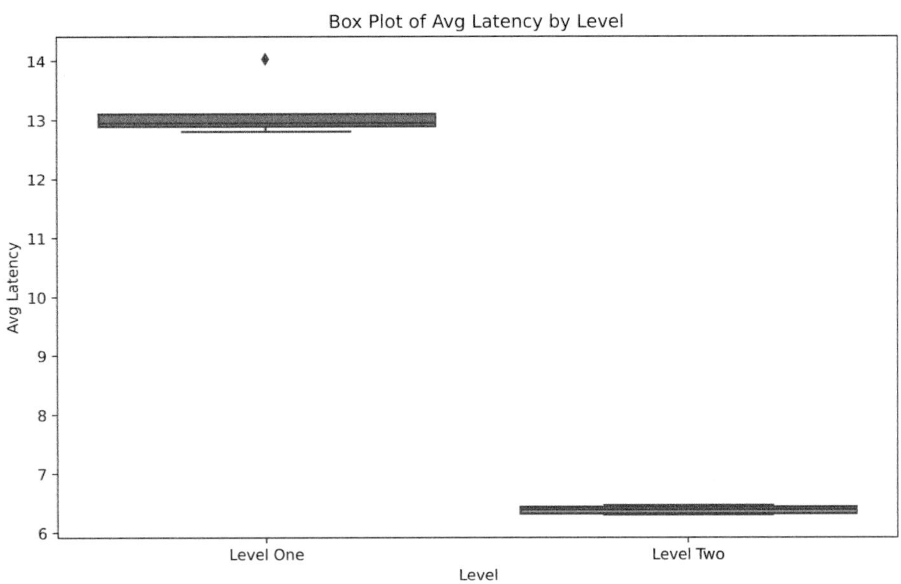

Figure 15.9 Reduced latency shown in a box plot.

process of separately representing each event provides the reader with a better understanding of the actual communication process of raft replication.

In Figure 15.9, "1" being the leader, accepts the client's appeal, and the event is marked by the letter "A." The command is sent to the hub to be broadcast to all available followers. In the next event 'B,' the hub transmits to all the remaining

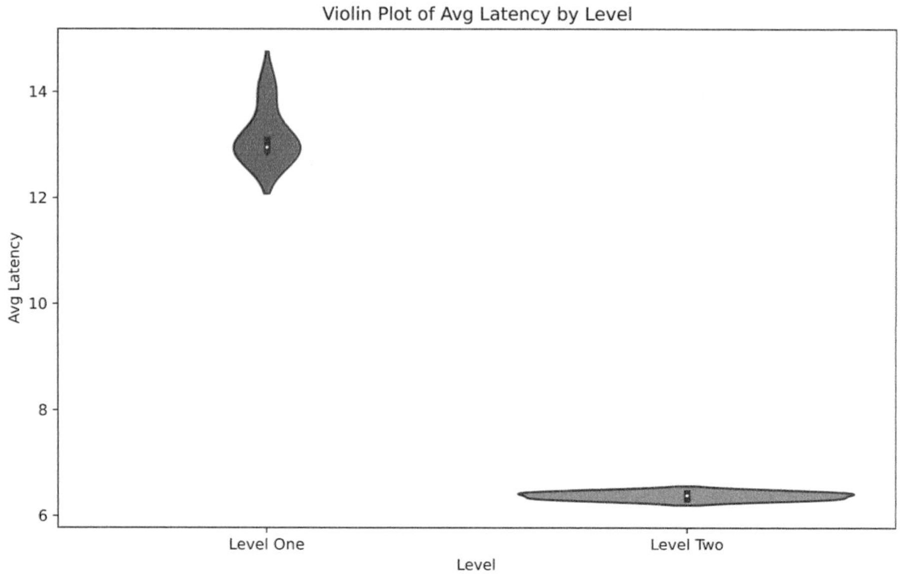

Figure 15.10 Comparison of average latency through violin plots.

nodes in the cluster simultaneously. Node "2" responds first and is indicated by the event "C." This response is propagated to the leader "1" through the hub. Since only one hub accepts all requests, acceptance of requests is always sequential. Hence, event "D" only happens after event' C.'

Followers 3, 4 and 5 responded to events D, E, and F through the central hub shown in Figure 15.10. The leader "1" finally responds positively to the client "C1," ending the completion of the replication cycle, generating integrity and reliability of the stored state machine command across all the nodes in the raft cluster.

Measured in terms of events, it took eight steps to complete a single "set" value from the client. This process is simple and guarantees consistency as shown in Figures 15.11 and 15.12; however, the leader is still the bottleneck, as it has to send and gather all requests and responses through a single hub.

Here, we illustrate how the proposed method changes the raft algorithm, wherein the process of broadcasting and accepting responses is delegated to a deputy. The following series of diagrams shows the same step-by-step event-driven process.

Instead of broadcasting through the hub, in the event A, the client request is only sent to the deputy leader, "2." This frees leader "1" to receive more client requests and delegate further. While the events D and E are still propagating, the leader '1' accepts another request from the client and starts to delegate to the proxy leader "2." While event E is still propagating to the proxy leader, that proxy leader also accepts event B1 from the leader. By the time of event F, the proxy leader broadcasts event C1 to the raft follower.

While the proxy leader "2" is sending a response from the Followers to the leader through event G, the reply from the 5 in the form of D2 is already being

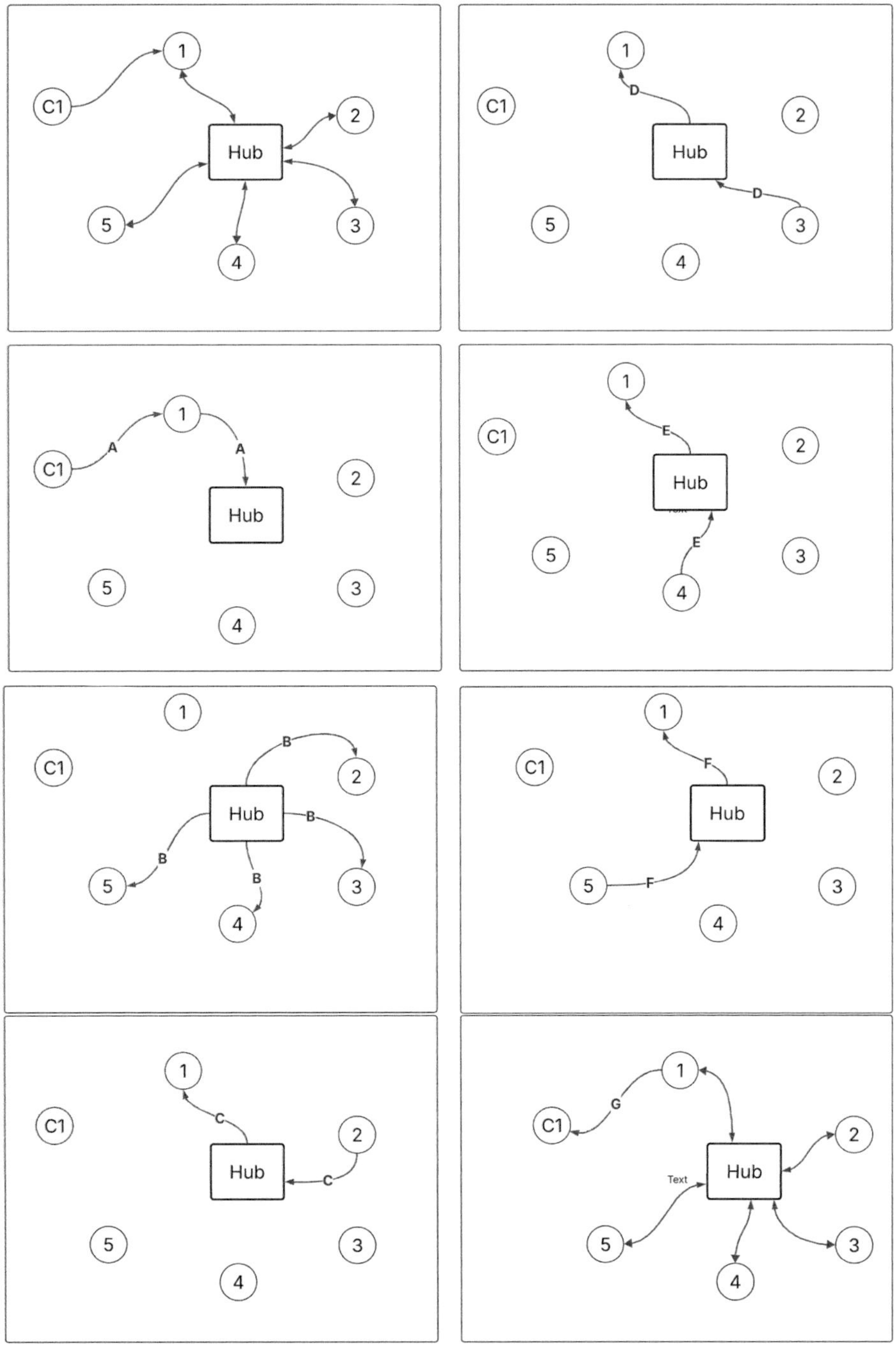

Figure 15.11 Client Command propagation through replication in a raft cluster.

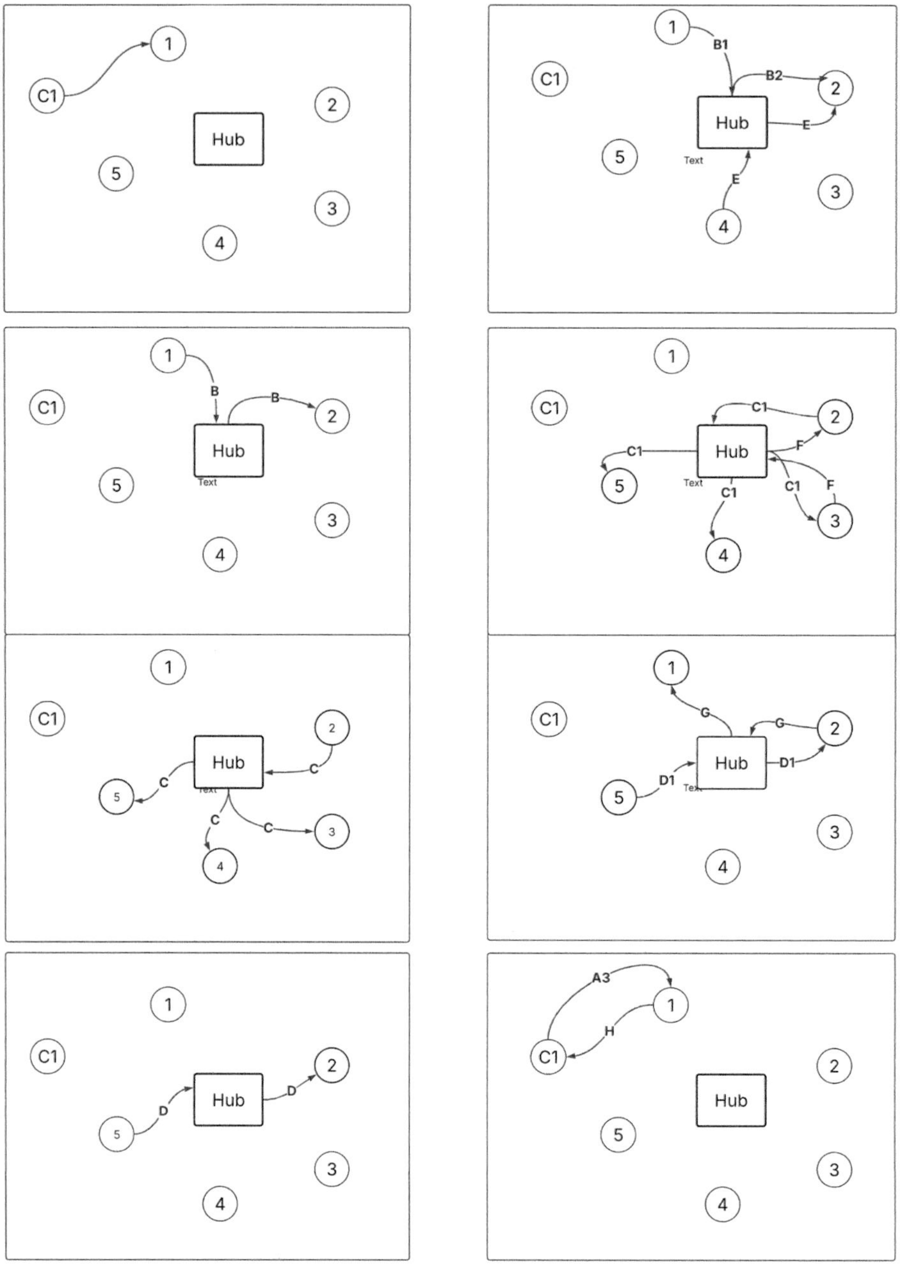

Figure 15.12 Client command propagation through replication in raft cluster.

received. In the last step, the client receives a response to the first request, whereas replication is halfway through for the second request.

The preceding graphical representation of sequential occurrences demonstrates the enhancement in which the adjusted raft system accommodates an increasing number of client solicitations and assigns them to a proxy figurehead. This strategy effectively distributes the communication workload in parallel without overwhelming the central focal point. Upon the arrival of the second client request to leader '1,' the adapted raft protocol had already acquired responses from the subordinates.

6. Conclusion and Future Studies

In this article, we tried to enhance the efficiency of the Raft consensus algorithm regarding throughput and scalability, aiming to enhance the comprehensive performance of a permissioned blockchain network employing Raft. Due to the important role of consensus within any blockchain network, enhancements to the consensus mechanism could lead to substantial performance enhancements based on the quality of the protocol's design. It has been demonstrated that the overall performance can be enhanced even with an increase in the quantity of nodes through the implementation of 2-level log replication broadcasting. The log replication component of the Raft algorithm was executed and subjected to simulation to authenticate the efficacy of the proposed concept. Significant enhancements in the mean throughput and a noticeable reduction in average latency were observed in a 25-node cluster. Further, we would like to evaluate a Raft cluster size of 50, 75, and potentially 100 nodes, as Hyperledger Fabric offers compatibility with numerous nodes.

References

[1] Ongaro, D. and Ousterhout, J. (2014). In Search of an Understandable Consensus Algorithm. 18.

[2] Xiao, Y., Zhang, N., Lou, W. and Hou, Y.T. (2020). A survey of distributed consensus protocols for blockchain networks. IEEE Commun Surv Tutor. 22: 1432–1465.

[3] Ye, Z., Tong, X., Fan, W., Zhang, Z. and Jin, C. (2023). A raft variant for permissioned blockchain. Int Conf Database Syst Adv Appl. 685–689.

[4] Yu, Y., Manolios, P. and Lamport, L. (1999). Model Checking TLA+ Specifications. pp. 54–66. *In*: Pierre, L. and Kropf, T. (eds.). Correct Hardware Design and Verification Methods. Berlin, Heidelberg: Springer. (Lecture Notes in Computer Science).

[5] Weizho Tang, Peiyao Sheng, Pronoy Roy, Xuechao Wang, Giulia Fanti and Pramod Viswanath. (2023). Raft-Forensics: High Performance CFT Consensus with Accountability for Byzantine Faults. DOI:10.48550/arXiv.2305.09123.

[6] Huang, D., Ma, X. and Zhang, S. (2018). Performance analysis of the raft consensus algorithm for private blockchains. IEEE Trans Syst Man Cybern Syst. 50: 172–181.

[7] Digital Equipment Corporation, Lamport L. (2019). The part-time parliament. *In*: Malkhi, D. (ed.). Concurrency: the Works of Leslie Lamport [Internet]. Association for Computing Machinery; 2019 [cited 2022 Oct 7]. Available from: https://dl.acm.org/citation.cfm?id=3335939.

[8] Lamport, L. (2006). Fast paxos. Distrib Comput. 19: 79–103.

[9] Lamport, L. (2005). Generalized consensus and Paxos.

[10] Sutra, P. and Shapiro, M. (2011). Fast genuine generalized consensus. 2011 IEEE 30th Int Symp Reliab Distrib Syst. 255–264.

[11] Marandi, P.J., Primi, M., Schiper, N. and Pedone, F. (2010). Ring Paxos: A high-throughput atomic broadcast protocol. 2010 IEEEIFIP Int Conf Dependable Syst Netw DSN. 527–536.

[12] Marandi, P.J., Primi, M. and Pedone, F. (2012). Multi-ring paxos. IEEEIFIP Int Conf Dependable Syst Netw DSN 2012: 1–12.

[13] Yanhua Mao, Flavio P. Junqueira and Keith Marzullo. (2008). Mencius: Building Efficient Replicated State Machines for WANs. 8th USENIX symposium on operating systems design and Implementation.

[14] Biely, M., Milosevic, Z., Santos, N. and Schiper, A. (2012). S-paxos: Offloading the leader for high throughput state machine replication. 2012 IEEE 31st Symp Reliab Distrib Syst. 111–120.

[15] Howard, H. (2014). ARC: analysis of Raft consensus.

[16] Nagawiecki, A. and Patterson, S. (2019). Layered Paxos: A Hierarchical Approach to Consensus.

[17] Whittaker, M. (2021). Compartmentalizing State Machine Replication.

[18] Savithri, M., Pradeepa, M. et al. (2023). Enhancement of QoS in Internet of Things Wearable Devices Dependent on 5G Technology. Wireless Communication and Mobile Computing 2023.

[19] Vivek Anand, M., Dr. Munish Sabharwal and Dr. Vijayalakshmi, S. (2022). Alleviating data storage challenge through virtualization of blockchain embedded with internet of things. Indian Journal of Computer Science and Engineering 13(2): 312–323, Doi:10.21817/Indjcse/2022/V13i2/221302005.

[20] Anand, M.V., Sabharwal, M. and Vijayalakshmi, S. (2022). Security intensification using blockchain coupled with internet of things: proposal, challenges and anatomization. 2022 4th International Conference on Advances in Computing, Communication Control and Networking (ICAC3N), Greater Noida, India, pp. 2044–2048, doi: 10.1109/ICAC3N56670.2022.10073978.

Chapter 16

Graph Partitioning Techniques for Ameliorating Compression of 3D LiDAR Point Cloud Data

S. Lakshmi Bala[1] and *P.L. Chithra*[2,*]

1. Introduction

The ability to collect extremely comprehensive three-dimensional (3D) spatial data has allowed LiDAR (Light Detection and Ranging) technology to completely transform a number of sectors. In order to collect precise and thorough environmental data, LiDAR systems are essential for anything from autonomous driving to urban planning. However, processing, transmission, and storage of the massive volume of data produced by LiDAR sensors are severely hampered. The need for effective compression methods to handle the massive amount of Point Cloud Data (PCD) has increased with the spread of LiDAR applications. Conventional compression techniques might not be adequate to handle the specific properties and needs of LiDAR point cloud data; it would be necessary to investigate more specialised techniques designed for this field.

Specifically for 3D LiDAR point cloud data, this paper provides a unique graph-based compression technique. The method uses graph partitioning techniques to effectively divide the point cloud into smaller fragments while maintaining important geometric characteristics and spatial linkages, demonstrating that the proposed approach effectively reduces data size while preserving data veracity and integrity through methodical examination and analysis.

[1] Research Scholar, Department of Computer Science, University of Madras.
[2] Professor, Department of Computer Science, University of Madras.
Email: lakshmibala.sj@gmail.com
* Corresponding author: chitrasp2001@yahoo.com

DOI: 10.1201/9781003635178-22

Further, effective compression methods become essential as the requirement for real-time LiDAR data processing and transmission increases across a range of applications, including robotics, urban mapping, and environmental monitoring. The proposed GPCLPCD compression technique not only addresses storage and transmission limits but also speeds up data processing and analysis by minimising the data footprint without sacrificing the accuracy and reliability of the reconstructed point cloud data. In order to improve the scalability, accessibility, and usefulness of LiDAR technology across a broad range of applications and sectors, the proposed work seeks to further LiDAR data compression approaches.

The importance of LiDAR technology and the difficulties presented by high PCD volumes are discussed in Section 1, Introduction. Examine the body of research on PCD compression to identify gaps and provide guidance for the creation of the proposed framework, as covered in Section 2. The framework and technical strategy for compressing LiDAR PCD, with a particular emphasis on graph-based compression methods, are covered in Section 3. Section 4 presents the results of the evaluation of the proposed compression framework, including quantitative measurements and a comparison to other approaches. Outline future research directions to improve LiDAR data compression techniques, and summarize the main discoveries discussed in Section 5.

2. Related Work

Smarter cities come from the advancement of technology. Autonomous driving (AV) is a crucial component of autonomous driving (AV), which is the primary component of smart cities. But the worth of life and human safety might also be impacted by the susceptibility of uncrewed vehicles [3]. For both serial and parallel computers, algorithms that effectively partition extremely irregular graphs and unstructured meshes are essential for creating effective solutions for solving a variety of problems across numerous application domains [1]. METIS routinely generates partitions that are 10% to 50% better than those generated by spectral partitioning methods [2]. Every point in the cloud has a 3D position and colour assigned to it. Based on a hierarchical transform, we develop a technique to compress the colours in point clouds, which is described in [4].

The image at full resolution is shown at the base of the pyramid, while the image at its lowest resolution is represented at its topmost layer. Between the pyramid's subsequent tiers, resolutions are doubled in both the horizontal and vertical dimensions, as explained [5]. Significant progress has been achieved in the development and assessment of algorithms for balanced (hyper)graph partitioning [6]. The foundation of robotics and computer vision, with potential uses ranging from object detection to scene analysis, is 3D LiDAR point cloud data [7]. In addition to accelerating later processing stages, this compression facilitates data storage and transfer [9]. By building feature graphs and enhancing the aggregation of the node-level representations, EGC2 is able to capture the link between the features of various nodes, as explained [10]. The goal is to select half of each partite's vertices in order to maximise the weight of all the edges that connect those selected vertices are discussed [11]. One of the most important functions of an autonomous car is the ability to locate and range obstacles. In recent years, several autonomous car systems have been shown. Detection and Ranging Light (LIDAR) sensors are among the most used stereo vision technologies. LIDAR is a detection and ranging technology that uses lasers and has features similar to those of radar [12]. The graph's edges, nodes, and node significance are all known prior to the transport problem being formulated. The dual characterization of the transport formulation, which can be set up for directed or undirected graphs, is expressed in terms of distributions over the nodes and is discussed [13].

The suggested method breaks a graph into a collection of large cliques, which are then used to condense and concisely depict the graph [14]. Graph partitioning finds applications mostly in numerical analysis, image segmentation, program mapping into parallel systems, and VLSI design. Over the previous forty years are discussed in [15]. An effective graph partitioning method that simultaneously achieves good load balancing among computers and minimal communication costs [16]. In order to calculate the precise edge weights for partitioning, current approaches require counting every instance of the motif. On the other hand, the enumeration becomes unaffordable for huge graphs [17]. Create graph partitioning techniques that automatically classify sewer network segments into compartments that are semi-distributed in [18]. 3D airborne LiDAR PCD compression using a deep learning model. The optimisation functions of Stochastic Gradient Descent and Mean Squared Error improve the quality of the decompressed image discussed in [19]. Using the circular differential cosine transformation and axis outlier identification techniques, the raw LiDAR data is cleaned and normalised in point cloud data discussed in [20].

3. Methodology

This work presents a novel graph-based compression system designed especially for LiDAR (Light Detection and Ranging) Point Cloud Data (PCD) compression. Graph partitioning techniques are used to divide the point cloud into manageable portions while maintaining important geometric characteristics and spatial linkages. This is an essential concept underlying the GPCLPCD approach.

The below Figure 16.1 is the methodology process composed of multiple discrete steps. The point cloud is first represented as a graph, with each node representing a data point and edges capturing the geographical proximity relationships between points. Using the state-of-the-art METIS graph partitioning method comes next. Dividing the graph into several partitions in order to minimise inter-partition connection and achieve a balance in partition sizes.

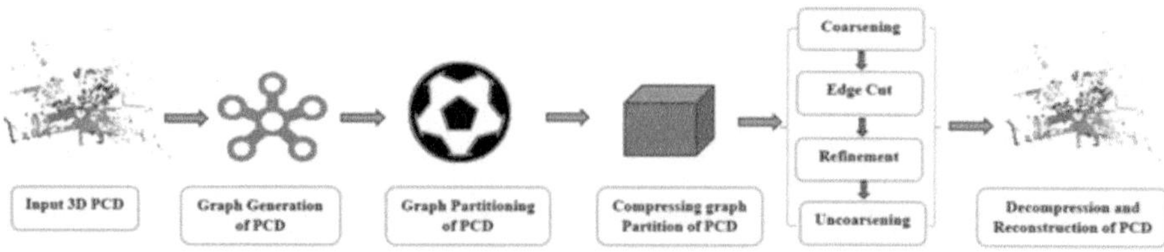

Figure 16.1 Block diagram of the proposed GPCLPCD method.

3.1 Initial Processing of Point Cloud Data

Reading the input Point Cloud Data is the initial step in the compression process (PCD). The position of each point in the PCD in three dimensions is indicated by its XYZ coordinates. The sample input 3D datasets are displayed in Figure 16.2.

$$P = \{p_1, p_2, \dots, p_n\} \tag{1}$$

Where, the points that were extracted out of the input PCD file are denoted by P. The total number of points in the point cloud is represented by n.

Figure 16.2 displays the process of GPCLPCD compression, as described in further steps.

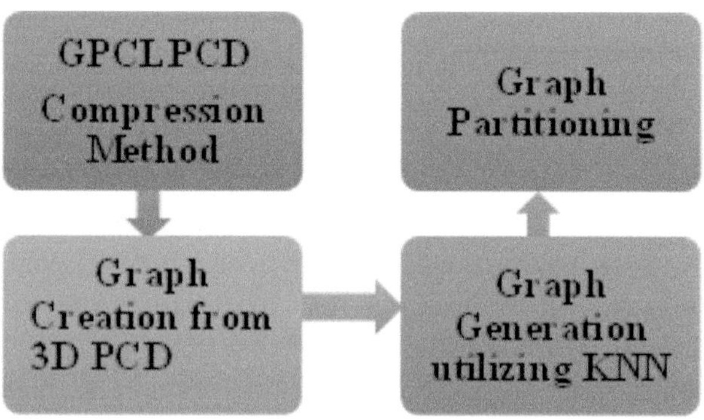

Figure 16.2 Process flow of the GPCLPCD method.

3.2 Graph Generation of Point Cloud Data

After obtaining point cloud data, a graph must be conceptualised in which nodes represent the points in the point cloud and edges represent the spatial interactions between them. Geometric attributes and relationships are computed more effectively from the graph-based form, which also makes it easier to analyse the point cloud data.

A node in the graph represents each point in the three-dimensional point cloud.

$$v_i = (x_i, y_i, z_i) \tag{2}$$

Where the corresponding coordinates of the point in three-dimensional space are x_i, y_i and z_i.

Distance measures like the Euclidean distance are used to identify the spatial proximity between points in the PCD. The calculation of the Euclidean distance $d_{i,j}$ between nodes v_i and v_j.

$$d_{i,j} = \sqrt{(x_j - x_i)^2 + (y_j - y_i)^2 + (z_j - z_i)^2} \tag{3}$$

Spatial links between points in the point cloud are represented by the edges connecting the graph nodes. From equation (3), if nodes v_i and v_j in the point cloud are spatially connected, then an edge (v_i, v_j) exists.

$$(v_i, v_j) \in E \tag{4}$$

If there is a spatial relationship between v_i and v_j. A set E consists of node pairs connected by edges. Every edge (v_i, v_j) in E represents a spatial link between the respective nodes in the graph v_i and v_j. The graph generation of 3D PCD is shown in Figure 16.3.

3.3 Utilizing K-Nearest Neighbour

Based on spatial proximity, the K-Nearest Neighbour (KNN) algorithm is used to determine each node's k nearest neighbours (v_i) in the graph. The KNN algorithm finds the k nodes in the 3D point cloud that are closest to each node (v_i) in terms of Euclidean distance. The neighbourhood N_i of node (v_i) is formed by these k closest neighbours, designated as:

$$N_i = \{v_{i1}, v_{i2}, \ldots, v_{ik}\} \tag{5}$$

Following the identification of each node k nearest neighbours, edges are created to link each node to its nearest neighbours. For each node v_i, an edge $(v_i, v_{i,j})$ for each nearest neighbour $v_{i,j}$ is constructed.

Therefore, for $1 \leq j \leq k$ an edge $(v_i, v_{i,j})$ prevails in the set of Edges E:

$$E = \{(v_i, v_{i,j}) \mid 1 \leq j \leq k\} \tag{6}$$

These edges show the spatial interactions in the 3D point cloud between the nodes v_i and its k nearest neighbours.

3.4 Graph Partitioning

The crucial step in GPCLPCD is graph partitioning. It entails creating smaller, more balanced subgraphs from the graph generated from the point cloud. This separation makes processing and storing data more efficient and lowers the computational complexity. The objective is to maintain about equal numbers of nodes in each partition while minimising the number of connections (edge cuts) that connect nodes in different partitions.

The graph created from the PCD is represented by

$$G = (V, E) \tag{7}$$

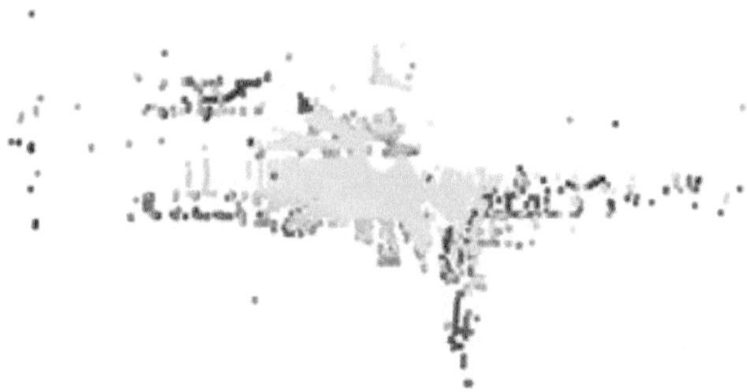

(a)

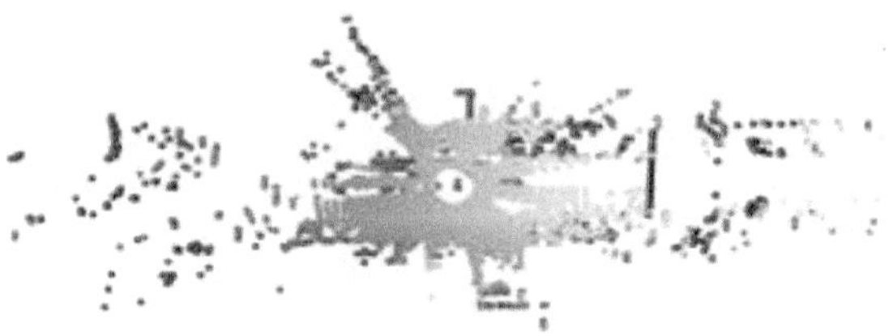

(b)

(c)

(d)

Figure 16.3 Sample Input 3D PCD images (a) scan0.pcd, (b) scan270.pcd, (c) scan2299.pcd and (d) scan25322.pcd.

Where the set of nodes (points in the point cloud) is denoted by $V \cdot E$ is the collection of edges that illustrate how the different points are arranged in space.

A list of adjacencies is utilised to depict the graph. The adjacency list holds a list of vertices that are next to each vertex v_i in range V. Formally, the adjacency list for a graph G with n vertices is expressed as:

The adjacency list represents the G with each element $Adj[i]$ is the adjacency list representation of v_i.

$$Adj[i] = \{v_j \in V \,|(v_i, v_j) \in E\} \;\forall\; v_i \;\in V \tag{8}$$

The objective is to divide G into k disjoint subsets $v_1, v_2, \ldots, v_k$ so that every subset has the same number of nodes and there are as few edges as possible between various subsets.

Reduce the number of edges between distinct divisions, or the edge cut.

$$Edge\ cut = \sum_{(v_i, v_j) \in E} \delta(part[i], part[j]) \tag{9}$$

Where $part[i]$ denotes the vertex v_i and δ is the indicator function.

$$\delta(a, b) = \{1\ if\, a \neq b\ 0\ if\, a = b \tag{10}$$

The balance constraint is expressed as:

$$|V_i| \approx \frac{|V|}{k} \forall i \tag{11}$$

3.5 Compression of Partitioning

Partitioning the graph representing the point cloud data into smaller, easier-to-manage segments is a critical step in the compression process that is accomplished with METIS. This division makes it possible to process individual partitions in parallel and compress them, which results in effective data transmission and storage.

A number of phases, including coarsening, refining, uncoarsening, and partitioning, are involved in the METIS compression process. Every stage has a distinct function in maximising the graph's partitioning that represents the point cloud data.

Coarsening

Aggregating nodes in the graph to minimise its size while maintaining its fundamental structure is known as coarsening. A coarser approximation of the original graph is produced by repeatedly collapsing nodes and the edges that surround them into super nodes. While maintaining crucial connectivity information, coarsening aids in lowering the partitioning GPCLPCD computational complexity.

$$Supernode = \{v_1, v_2, \ldots, v_k\}$$

Where the initial nodes that are combined to produce the supernode are represented by the $v_1, v_2, \ldots, v_k$.

$$Edge(Supernode_1, Supernode_2) = \sum_{i=1}^{k_1} \sum_{j=1}^{k_2} Edge(v_i, v_j) \tag{12}$$

Where k_1 and k_2 are the numbers of nodes in $Supernode_1$, $Supernode_2$ respectively. The *Edge* (v_i, v_j) indicate the edge weight between the nodes v_i and v_j in original graph.

The graph G is iteratively coarsened into a sequence of smaller graphs $G_0, G_1, \ldots, G_m$ so that each G_{i+1} is a contracted form of G_i. The graph size of PCD is decreased through coarsening while maintaining its structure. The coarsening can be expressed as:

$$G_{i+1} = contract\ (G_i) \tag{13}$$

Refinement

The goal of refinement is to raise the calibre of the partitioning that results from coarsening. In order to reduce edge, cut and enhance partition balance, iteratively fine-tune the partitioning by shifting nodes between partitions. Refinement aids in the partitioning solution's optimisation and improvement in relation to particular goals, such as reducing communication costs or maximising computational efficiency.

$$Edge\ cut(P) = \sum_{i \in V} \sum_{j \in V} w(i, j) \cdot \delta(P_i, P_j) \tag{14}$$

Where P represents the graph partitioning, V is the graph set of nodes, and $w(i, j)$ is the weight of the edge between nodes i, and j. An indicator function called $\delta(P_i, P_j)$ equals 1 when nodes i and j are in different partitions and 0 otherwise.

Uncoarsening

By gradually fine-tuning the partitioning solution acquired on the coarsened graph, uncoarsening undermines the coarsening process. In this step, the partitioning solution is expanded to the size of the original graph while preserving its quality. The process of uncoarsening guarantees that the partitioning solution that is ultimately chosen faithfully captures the original point cloud data connectedness and structure.

Uncoarsening is the procedure by which the partitions identified in the coarsest graph are transformed back to the original graph. The partitions are balanced, and the edge cut is reduced at each stage of refinement.

$$P_{i-1} = refine\ (P_i, G_{i-1}) \tag{15}$$

In order to enhance the partition, the edge cut and balance during uncoarsening, METIS refines the partitions utilising local search as an optimization technique.

Partitioning

The process of dividing a graph into several partitions based on a refined partitioning solution acquired through the steps of coarsening, refining, and uncoarsening is known as partitioning. The aim is to achieve a balanced distribution of nodes across partitions while minimising the edge cut. Each partition represents a fraction of the original point cloud data. Partitioning the compressed data into smaller, more manageable chunks allows for more effective processing, transport, and storage.

To get an initial partition, a partitioning step is applied to the smallest graph G_m.

$$P_m = partition\ (G_m) \tag{16}$$

3.6 Decompression and Reconstruction of Point Cloud Data

Expanding the compressed partitions back to their original form and reconstructing the entire point cloud are the processes involved in decompressing and reconstructing point cloud data. The goal of this procedure is to minimise data loss during compression while recovering the original spatial information.

When graph hierarchical compression techniques, such as METIS, are used to compress point cloud data, decompression and reconstruction are essential steps in recovering the original PCD. While the proposed GPCLPCD method is a lossless technique, ensures that every aspect and feature of the original dataset is preserved during the compression process.

4. Experimental Results

The results of applying the proposed GPCLPCD compression methodology to a Sydney urban dataset [8] are presented in this experimental results section. In-depth analysis and measurements are provided in this part to assess the proposed method efficacy and efficiency. The outcomes are displayed in terms of edge cuts, partition balance, compression ratios, and reconstruction quality (Hausdorff distance and time required for compression. Table 16.1 below depicts the performance of the proposed work. First the input is the 3D PCD file. The total number of points in the input PCD file is represented by this metric. Every point in the PCD becomes a node in the graph after the point cloud is transformed into a graph. The graph edges are formed according to the spatial connections between the points. The result is a collection of nodes divided into several divisions with the goal of minimising edge cuts across partitions and balancing the number of nodes in each partition.

Table 16.1 Performance of proposed GPCLPCD method.

Input 3D PCD	Number of Points in PCD	Number of Nodes in Graph	Number of Edges in Graph	Partitioning in Nodes				Compression Ratio	Compression Time
Scan0.pcd	42,165	42,165	243,658	10,540	10,542	10,541	10,542	2.08	0.79
Scan270.pcd	56,283	56,283	287,605	14,082	14,059	14,072	14,070	2.10	0.78
Scan10974.pcd	55,714	55,714	288,609	13,928	13,929	13,932	13,925	2.09	0.77
Scan11290.pcd	57,188	57,188	295,820	14,294	14,290	14,295	14,309	2.08	0.99
Scan2299.pcd	55,107	55,107	282,721	13,772	13,781	13,777	13,777	2.08	0.76
Scan2738.pcd	59,395	59,395	304,694	14,848	14,849	14,848	14,848	2.09	0.83
Scan25322.pcd	55,302	55,302	287,538	13,824	13,822	13,827	13,829	2.07	0.75

4.1 Datasets used for GPCLPCD

To evaluate the performance of the proposed compression approach using real Sydney urban point cloud datasets [8]. The datasets covered a range of 3D construction types and differed in size and complexity.

4.2 Measures of Compression Performance

The compression performance was assessed using the following metrics:

Compression Ratio

One of the most important metrics for assessing a compression algorithm's effectiveness is the compression ratio, or CR. It measures the amount of data compressed by the compression process. Since the compressed data is substantially less than the original data, a higher compression ratio denotes a more effective compression.

The ratio of the original data size to the compressed data size is the compression ratio (CR) as shown in Table 16.2:

$$CR = \frac{S_{original}}{S_{compressed}} \tag{17}$$

The size of the original point cloud data, expressed in Kilo bytes, is $S_{original}$. The compressed point cloud data file size, expressed in kilobytes, is $S_{compressed}$.

Table 16.2 Outcomes of Compression ratio for Individual Partitioning Metrics.

Partition	Scan 270.pcd			
	No. of Nodes	**Original size (Bytes)**	**Compressed size (Bytes)**	**Compression Ratio**
0	14,082	760,676	165,112	4.61
1	14,059	759,434	162,175	4.68
2	14,072	760,136	160,692	4.73
3	14,070	760,028	154,843	4.91
TOTAL	56,283	3,040,274	642,822	4.73

Partition Balance

An approach to partitioning that was balanced was shown by the roughly equal number of nodes in each partition. Compression is more effective when the graph is divided into smaller subgraphs. Comparing compressing a single huge graph to compressing smaller subgraphs allows for a more densely packed and lower overhead. The compressed sizes shown in Table 16.3 represent the efficiency for scan270.pcd.

The compression GPCLPCD algorithm more efficiently takes advantage of local patterns and redundancies inside each division by dividing the graph into smaller chunks. The quality of the partitioning can have a direct impact on the compression ratios displayed in Table 16.1.

Table 16.3 Summary of Partitioning Results for Scan270.pcd.

Partition	Scan 270.pcd				
	No. of Nodes	**Partition Balance**	**Edge Cuts**	**Original size (Bytes)**	**Compressed size (Bytes)**
0	14,082	23	147	760,676	165,112
1	14,059	25	152	759,434	162,175
2	14,072	20	149	760,136	160,692
3	14,070	21	150	760,028	154,843
TOTAL	56,283	111	741	3,040,274	642,822

Edge cut

Information on the graph-based splitting of a point cloud dataset called "Scan 270. pcd" is displayed in Table 16.3. The number of Nodes shows how many nodes (points) are present in each partition. The partition balance is a measure of how evenly nodes are spread across partitions; it is the difference between the maximum and smallest number of nodes among partitions. Determine the overall efficacy of the partitioning technique in spreading the data equitably, and draw attention to any notable disparities in the number of nodes between partitions. The quantity of edges that are cut during the partitioning process indicates the degree of disruption to connections that the partitioning causes.

The importance of edge cuts in preserving connectivity in the dataset of PCD in Table 16.3. Edge cuts represent the number of connections between nodes in various partitions. Aggregated data from all partitions, such as the total number of nodes, total edge cuts, and total original and compressed sizes, is provided by the "TOTAL" row. This table aids in evaluating how well the partitioning procedure balanced the partitions, maintained connectivity, and produced compression.

Hausdorff Distance

A metric used to quantify how similar two sets of points are to one another in a metric space is the Hausdorff distance. By taking into account the greatest distance from any point in one set to the nearest point in the other set, it quantifies the dissimilarity of two points in the PCD.

In a metric space, the Hausdorff distance $H(X,Y)$ between two-point sets (X,Y) is expressed as:

$$H(X,Y)(max_{x\in X}\, min_{y\in Y}\, \|x-y\|,\, max_{y\in Y}\, min_{x\in X}\, \|x-y\|) \tag{18}$$

Where X and Y are the two-point sets, and $\|x-y\|$ depicts the distance between x and y. The *min* and *max* show the maximum and minimum value. The expression $max_{x\in X}\, min_{y\in Y}\, \|x-y\|$ determines the maximum distance between any point in Y and its closest neighbour in X. Similarly, $max_{y\in Y}\, min_{x\in X}\, \|x-y\|)$ expression denotes the maximum distance between any point in X and its closest neighbour in Y. The greater of these two values is the Hausdorff distance. The below Table 16.4 displays the values of distance between PCD.

Table 16.4 Outcomes of Hausdorff distance.

PCD File	No. of Nodes	No. of Edges	Compression Ratio	Hausdorff distance
Scan0.pcd	42,165	243,658	2.08	2,372
Scan270.pcd	56,283	287,605	2.10	2,732
Scan10974.pcd	55,714	288,609	2.09	2,758
Scan11290.pcd	57,188	295,820	2.08	2,784
Scan2299.pcd	55,107	282,721	2.08	2,764

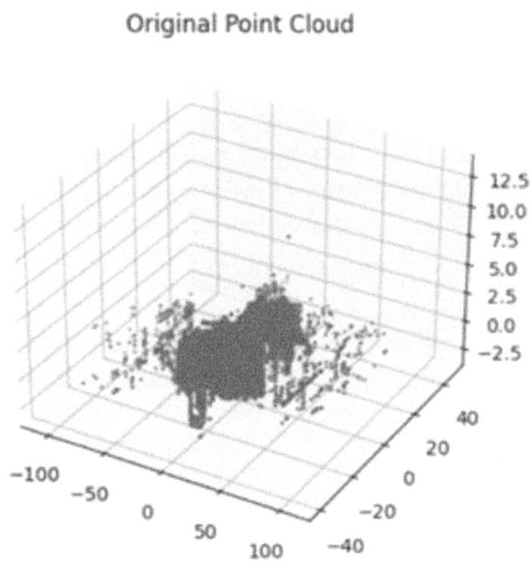

Figure 16.4 Raw 3D Input PCD (scan270.pcd).

The raw data is represented by the original point cloud. An overview of the density and structure of the dataset is given by Figure 16.4, which shows the spatial distribution of dots in three dimensions.

The first step in the graph generation procedure is to create a graph in the point cloud data. The process entails transforming the three-dimensional point cloud into a network of nodes and edges, wherein each node symbolises a point inside the cloud and the edges signify the connections among these points. Also, 56,283 points make up the point cloud in the above illustration, Figure 16.5. Every point in the point cloud is added to the graph as a node. This yields 56,283 nodes in the process.

Using the k-nearest neighbours (k-NN) technique, edges are created between the nodes. This method creates edges based on proximity by joining each point to its closest neighbours.

The graph produced by applying the k-nearest neighbours (KNN) algorithm to the point cloud data is shown in Figure 16.6. A point from the point cloud collection is represented by each node in the graph. Where each node is connected to its k nearest neighbours based on Euclidean distance, edges between nodes reflect spatial proximity.

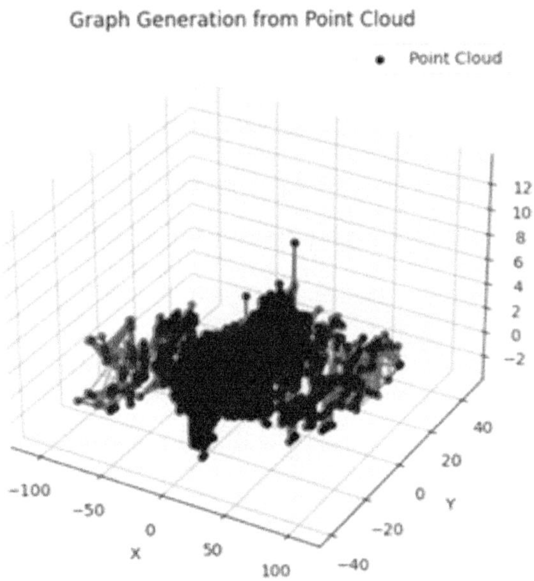

Figure 16.5 Graph generation of 3D PCD.

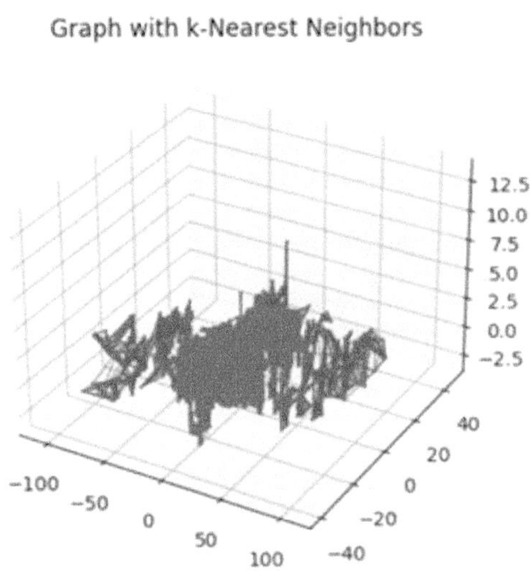

Figure 16.6 Graph with KNN of 3D PCD.

The structure of the scan270.pcd dataset and the connectivity between points are visualised in the graph, which is essential for establishing spatial relationships and patterns in the PCD.

The graph is partitioned into four separate partitions, each denoted by a different colour, as shown in Figure 16.7, output of the PCD. The METIS algorithm, which seeks to split the graph into balanced subgraphs while minimising edge cuts, is what led to the partitioning. The colour of each division makes it visually distinct, making

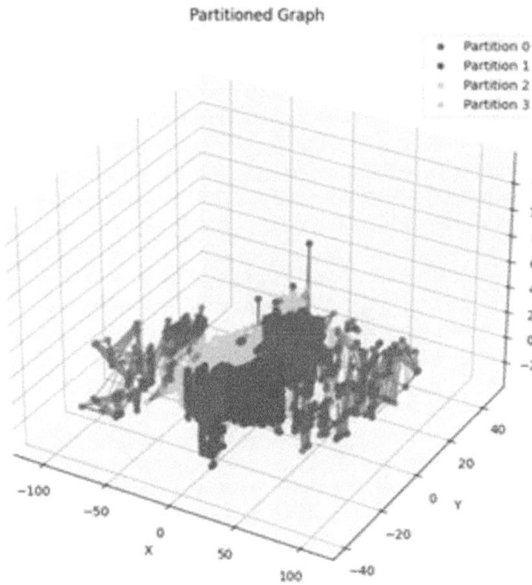

Figure 16.7 Partitioned Graph of PCD.

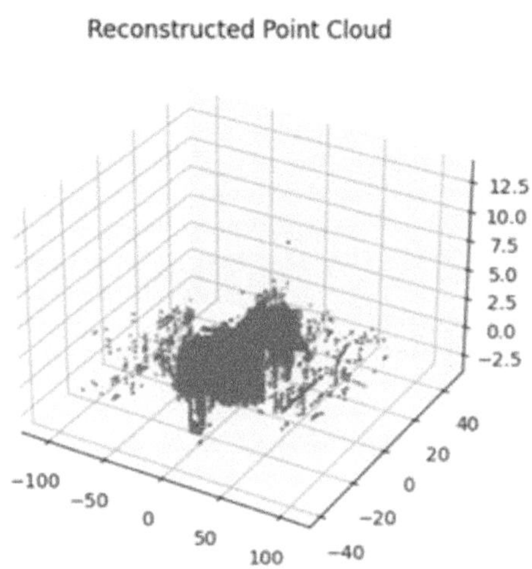

Figure 16.8 Reconstruction of the GPCLPCD method.

it easier to recognise the divided areas inside the graph. A clear visualisation of the partitioning structure by designating distinct colours to nodes that belong to different partitions. This partitioned GPCLPCD model divides the workload among several divisions, allowing for effective processing and analysis of the graph data.

Finally, Figure 16.8 visualise the PCD that has been reconstructed. The compressed partitions that were obtained during the graph partitioning phase are decompressed and then rebuilt to accomplish this reconstruction. The evaluation of

Table 16.5 Performance of proposed GPCLPCD with other compression techniques.

PCD File	File size of Original	File size of WinRAR	File size of LZ4 algorithm	Compression Ratio	File size of Proposed GLPCD
Scan0.pcd	824 KB	456 KB	606 KB	2.08	**396.15 KB**
Scan270.pcd	1,100 KB	619 KB	842 KB	2.10	**528.85 KB**
Scan10974.pcd	1,089 KB	620 KB	878 KB	2.09	**518.57 KB**
Scan11290.pcd	1,118 KB	625 KB	922 KB	2.08	**534.93 KB**
Scan2299.pcd	1,077 KB	608 KB	821 KB	2.08	**517.93 KB**
Scan2738.pcd	1,161 KB	645 KB	958 KB	2.09	**558.17 KB**

the fidelity of the compression-decompression process is achieved by comparing the original point cloud with its reconstructed equivalent using this visualisation.

The usefulness of the compression technique in maintaining the key characteristics and structural integrity of the original PCD is assessed through examining the rebuilt PCD.

Table 16.5 depicts the performance analysis of the proposed GPCLPCD technique, which outperforms well with WinRAR and with the LZ4 algorithm

5. Conclusion

Through the use of graph partitioning algorithms, the GPCLPCD method has proven to be effective at compressing massive amounts of point cloud data. The method preserved the original data quality and integrity while significantly reducing the amount of data. The PCD was transformed into a graph with 287,605 edges and 56,283 nodes, which successfully represented the spatial relationships among the points and provided a strong framework for compression and partitioning. A significant amount of equity in data distribution across partitions is shown by the partition balance of 23 nodes, which is essential for load balancing and parallel processing. With a compression ratio of 52%, the data size was significantly reduced from 1,100 KB to 660 KB, resulting in a reduction in storage needs. With 147 inter-partition edges, the edge cut metric represented the degree of connectedness between various partitions. The effectiveness of the GPCLPCD approach was proved by the average compression time of exactly 0.78 seconds and compared with WinRAR and the LZ4 algorithm. It is clear from the rebuilt point cloud that the geometric and spatial integrity of the data was not compromised throughout the compression process 56,283 points were retained.

This approach is appropriate for a number of applications that require large-scale 3D data compression, including geographic information systems (GIS), autonomous navigation, and virtual reality, due to its balanced data distribution and efficient processing time.

Subsequent research endeavours could emphasize refining the compression algorithms, discovering other partitioning techniques, and assessing the methodology across various datasets to confirm its resilience and adaptability.

References

[1] Karypis, G. (2011). METIS and ParMETIS. *In*: Padua, D. (eds.). Encyclopedia of Parallel Computing. Springer, Boston, MA. https://doi.org/10.1007/978-0-387-09766-4_500.

[2] METIS - Serial Graph Partitioning and Fill-reducing Matrix Ordering | Karypis Lab. glaros.dtc.umn.edu. http://glaros.dtc.umn.edu/gkhome/metis/metis/overview.

[3] Abbasi, R., Bashir, A.K., Alyamani, H.J., Amin, F., Doh, J. and Chen, J. (2022). Lidar point cloud compression, processing and learning for autonomous driving. IEEE Transactions on Intelligent Transportation Systems, pp. 1–18, doi: https://doi.org/10.1109/tits.2022.3167957.

[4] de Queiroz, R.L. and Chou, P.A. (2016). Compression of 3D point clouds using a region-adaptive hierarchical transform. IEEE Transactions on Image Processing 25(8): 3947–3956, Aug. 2016, Doi: https://doi.org/10.1109/tip.2016.2575005.

[5] Hierarchical Mode - an overview | ScienceDirect Topics. www.sciencedirect.com. https://www.sciencedirect.com/topics/computer-science/hierarchical-mode.

[6] Ümit Çatalyurek, Karen Devine, Marcelo Faraj, Lars Gottesburen, Tobias Heuer, Henning Meyerhenke et al. (2023). More recent advances in (hyper)graph partitioning. ACM Computing Surveys 55(12): 1–38, Mar. 2023, Doi: https://doi.org/10.1145/3571808.

[7] Chithra, P.L. and Lakshmi, S. (2024). Convolutional neural network model for stratification of LiDAR 3D point cloud data. Jan. 2024, Doi: https://doi.org/10.4108/eai.23-11-2023.2343218.

[8] A. C. for F. Robotics. Sydney Urban Objects Dataset. www.acfr.usyd.edu.au. https://www.acfr.usyd.edu.au/papers/SydneyUrbanObjectsDataset.shtml.

[9] Chithra, P.L. and Lakshmi bala, S. (2023). Deep Learning Model for 3D LiDAR Point Cloud Codec Based on Tensor. Nov. 2023, Doi: https://doi.org/10.1109/icccis60361.2023.10425292.

[10] Jinyin Chen, Haiyang Xiong, Haibin Zheng, Dunjie Zhang, Jian Zhang, Mingwei Jia et al. (2023). EGC2: Enhanced graph classification with easy graph compression. Information Sciences 629: 376–397, Jun. 2023, Doi: https://doi.org/10.1016/j.ins.2023.01.114.

[11] Li, G., Sun, J., Sun, Z., Du, D. and Zhang, X. (2024). An approximation algorithm for high-dimensional table compression on balanced K-partite graph. Computers & Electrical Engineering 113: 109048–109048, Jan. 2024, Doi: https://doi.org/10.1016/j.compeleceng.2023.109048.

[12] Kamalkumar, V., Sonali, S., Sindhuja, V. and Sriharish, A. (2022). LIDAR based self driving car. Jan. 2022, Doi: https://doi.org/10.4108/eai.14-5-2022.2318893.

[13] Garg, V. and Jaakkola, T. (2024). Solving graph compression via optimal transport. Accessed: Jun. 01, 2024. [Online]. Available: https://www.mit.edu/~vgarg/neurips2019compression.pdf.

[14] Rossi, R.A. and Zhou, R. (2018). GraphZIP: a clique-based sparse graph compression method. Journal of Big Data 5(1): Mar. 2018, Doi: https://doi.org/10.1186/s40537-018-0121-z.

[15] Graph Partitioning | Wiley. Wiley.com. https://www.wiley.com/en-us/Graph+Partitioning-p-978111860 1259 (accessed Jun. 01, 2024).

[16] Onizuka, M., Fujimori, T. and Shiokawa, H. (2017). Graph partitioning for distributed graph processing. Data Science and Engineering 2(1): 94–105, Feb. 2017, Doi: https://doi.org/10.1007/s41019-017-0034-4.

[17] Huang, S., Li, Y., Bao, Z. and Li, Z. (2021). Towards efficient motif-based graph partitioning: an adaptive sampling approach. Singapore Management University Institutional Knowledge (InK) (Singapore Management University) Apr. 2021, Doi: https://doi.org/10.1109/icde51399.2021.00052.

[18] Dobson, B., Watson-Hill, H., Samer Muhandes, Borup, M. and Mijic, A. (2022). A reduced complexity model with graph partitioning for rapid hydraulic assessment of sewer networks. 58(1) Jan. 2022, Doi: https://doi.org/10.1029/2021wr030778.

[19] Tamilmathi,A.C. and Chithra, P.L. (2021). Deep learned quantization-based codec for 3D Airborne LiDAR point cloud images. Frontiers in Robotics and AI 8, May 2021, Doi: https://doi.org/10.3389/frobt.2021.606770.

[20] Tamilmathi, A.C. and Chithra, P.L. (2022). Tensor block-wise singular value decomposition for 3D point cloud compression. Multimedia Tools and Applications 81(26): 37917–37938, Apr. 2022, Doi: https://doi.org/10.1007/s11042-021-11738-7.

Index

O

P

R

S

T

U

V

About the Editors

Dr. S. Vijayalakshmi is a Professor in the School of Sciences at CHRIST University, Pune Lavasa Campus, India. She has over 25 years of teaching experience, with research interests focused on Image Processing and the Internet of Things (IoT). She has contributed to Scopus, WoS and SCI-indexed journals, as well as numerous national and international conferences. She has also authored and edited books and chapters on topics such as artificial intelligence, IoT, deep learning, big data, and blockchain. Her work extends beyond publishing—she has filed and published **over 25 patents**, covering innovations like IoT-based health monitoring systems, smart assistive devices for the visually impaired, agricultural robots, and various smart IoT gadgets. On the research mentorship front, she has guided 4 Ph.D. scholars to completion and is currently supervising 3 more. She has also mentored over 500 students on academic projects. She has received research funding from bodies like CSIR, SERB, ICSSR and seed grants from CHRIST University. She is a Google Cloud Certified Engineer. She is an active member of professional bodies including IEEE, ACM, IFERP and IAENG and has been recognized with multiple **Best Paper** and **Best Presenter** awards at national and international conferences.

Dr. Savithri M. is an Assistant Professor in the School of Sciences at Christ University, Pune Lavasa Campus. She has over 20 years of teaching experience. Her research interests include IoT, artificial intelligence, data-driven applications, and network security. She has contributed to journal publications, book chapters, conference proceedings, and patents. She actively participates in conferences, collaborative research projects, and academic initiatives. Dr. Savithri is committed to teaching, mentoring students, and promoting innovation, interdisciplinary learning, and research-oriented education in Computer Science.

Dr. Lija Jacob is an Associate Professor and Head of the Department of Statistics and Data Science and Computer Science at Christ (Deemed to be University), Pune Lavasa Campus. With over 22 years of teaching and research experience, her work spans artificial intelligence, data science, and image processing. She has authored books, patents, and numerous research publications, and actively mentors young researchers while contributing to interdisciplinary innovation and technology-driven solutions for societal impact.